ANNUAL CROPPING SYSTEMS IN THE TROPICS

ANNUAL CROPPING SYSTEMS IN THE TROPICS

AN INTRODUCTION

M. J. T. NORMAN

A University of Florida Book

University Presses of Florida
Gainesville

Library of Congress Cataloging in Publication Data

Norman, Michael John Thornley.
Annual cropping systems in the tropics.

"A University of Florida book."
Includes bibliographies.
1. Cropping systems—Tropics. 2. Agriculture—Tropics. I. Title.
S604.N67 631.5'0913 79-10625
ISBN 0-8130-0632-5

The University Presses of Florida is the
scholarly publishing agency for the
State University System of Florida

Typography by Modern Typographers, Clearwater, Florida
Printed in U. S. A.

Contents

Preface

The conventional building blocks of university teaching in tropical agricultural science are much the same as those for agricultural science in general: crop agronomy, physiology and breeding, animal husbandry, nutrition and breeding, soil science, etc., within a tropical context. When in 1971 a graduate coursework program in tropical agronomy was developed in the Department of Agronomy and Horticultural Science, University of Sydney, the core curriculum—in this instance confined largely to the plant sciences—was organized along such conventional lines. However, it became clear that this alone was inadequate to reveal to students how tropical crop farming actually operates as a biological and physical system. As a consequence a short course on tropical farming systems was developed to provide a unifying element for the diverse agronomic curriculum.

In 1976, while on study leave at the Center for Tropical Agriculture, Institute of Food and Agricultural Sciences, University of Florida, the author was given the opportunity to amplify and present the course to graduate students, and during a second visit in 1977 to modify and repeat the course. It is from these successive steps that this volume has developed.

The development of a textbook from a course of lectures is a common progression in university science. In this instance there is perhaps more justification than usual, since to the author's knowledge there is no other publication that attempts to bring together in one volume the basic biological and physical principles underlying the pattern and operations of tropical farming systems.

The book is designed as an introduction to the subject. It is assumed that the reader has the background of a first degree in agricultural science and some acquaintance with tropical agriculture. The text makes no claim to review the very extensive literature of its field, and it is not supported by an exhaustive list of references. Rather, an attempt has been made to lead the reader to useful reviews and collections of papers that treat at the next level of detail of topics dealt with summarily in the book. Finally, although the subject is farming systems, the mode of approach is qualitative and the volume is not concerned with systems analysis in the mathematical sense.

Every writer on a broad topic, in the delineation of subject matter and choice of examples, reveals his personal experience and the boundaries of his competence. This book is restricted to annual cropping systems, since the author's association with perennial crop systems and pastoral systems—except for the atypical example of North Australia—has been limited. The geographical bias is toward Africa and Asia; tropical America is somewhat neglected for two reasons: the bulk of scientific literature on tropical farming systems is from Africa, and the author's experience is largely Asian.

The author would like to express his gratitude to Dr. Hugh Popenoe, Director of the Center for Tropical Agriculture, University of Florida, for providing the opportunity for the book to be written. The friendly tolerance of Dr. Michael Mullins, Professor of Horticulture, University of Sydney, toward the disruption of departmental administration occasioned by two study leave absences is gratefully acknowledged. Sincere thanks are also due to Drs. Coleman Ward, Gerald Mott, and Darrel McCloud, Department of Agronomy, University of Florida, for their support. I am deeply indebted to Mr. Robert Wetselaar, Division of Land Use Research, Commonwealth Scientific and Industrial Research Organisation, Canberra; to Dr. Richard Fluck, Department of Agricultural Engineering, University of Florida; and to Dr. Jacob Kampen, International Crops Research Institute for the Semi-Arid Tropics, Hyderabad, for their critical comments.

ANNUAL CROPPING SYSTEMS IN THE TROPICS

Chapter 1

What Is a Farming System?

■ 1.1 Introduction

■ 1.1.1 *Aim of the book.* The study of farming systems began within the discipline of geography. Agricultural geographers, searching for general relations between the observed patterns of farming in a region and the environmental and socioeconomic forces operating to mold such patterns, had of necessity to develop a taxonomic framework within which the individual and infinitely variable farming units could be fitted. From this work, farming system typologies developed, on the basis of which more intensive studies were made and continue to be made: for example, those of Duckham and Masefield (1971) and of Grigg (1974).

Agricultural economists have made an invaluable contribution to our appreciation of farming systems. Since those concerned with the microeconomics of production need to evaluate all the agricultural operations of the farming units that they examine, they are perhaps more likely than others to see the whole picture and to develop an understanding of how farming systems operate. With respect to the tropics, the work of Boserup (1965), Clark and Haswell (1970), Haswell (1973), and Ruthenberg (1971) amply attests to this.

At a more detailed level, since their field approach to fact-finding tends to be on a micro scale, social anthropologists have also made substantial additions to our knowledge of the farming systems of less advanced societies. The work of Brookfield and Brown (1963) illustrates how effective can be the joint approach of a geographer and an anthropologist.

A new method of approach is now established: the application with the aid of the computer of the techniques of simulation modeling to the biological, physical, and economic processes of farming systems. Books and journals concerned with the modeling of agricultural systems and subsystems are appearing: for instance Dent and Anderson (1971) and the journals *Agricultural Systems* and *Agro-Ecosystems*. It is clear that we are at present harvesting only the very earliest fruits of an approach that will have profound effects on our comprehension of how farming systems operate.

However, the contribution to farming systems research of the biological and physical disciplines of agricultural science—agronomy, soil science, and animal science—has until recently been disappointingly small. Trained in a reductionist approach to problem-solving, research workers in these fields have in the past been content to isolate small facets of agricultural processes for their attention. There are, of course, noticeable exceptions: the work of Nye and Greenland (1960) on shifting cultivation, for example.

Happily the situation is changing, and the book by Spedding (1975) is illustrative of recent developments. Furthermore, the change has been most marked where the need is greatest: in relation to the developing world. There the creation of multidisciplinary teams of research workers collectively engaged in research on farming systems as a whole is a most welcome trend: examples include programs at the International Rice Research Institute (IRRI), the International Institute for Tropical Agriculture (IITA), and the International Crops Research Institute for the Semi-Arid Tropics (ICRISAT), in the Philippines, Nigeria, and India respectively.

The present volume seeks in a modest way to contribute to a continued redressing of the balance. Written by an agronomist, it is primarily concerned with the biological and physical processes operating in tropical farming systems, while at the same time it attempts to relate their operation to the socioeconomic context of the farming unit. Reflecting the limits of the author's experience, the book is confined to tropical farming systems in which annual cropping predominates.

■ 1.1.2 *Definition of a farming system.* If a representative group of agricultural scientists, together with geographers, ecologists, anthropologists, economists, and systems analysts concerned with agriculture, were asked to define what they understood by the terms "agricultural system" or "farming system," it is likely that they would provide a heterogeneous set of answers. Perhaps they could all agree that there is no essential difference between the two terms—and in this book farming system is preferred—but it is probable that they would agree on little else.

Surprisingly, the authoritative general texts on farming systems—Duckham and Masefield (1971), Ruthenberg (1971), or Grigg (1974)—make little or no effort to define their subject. To fill this gap, two strongly contrasting definitions are given below:

First, from an anthropologist (de Schlippe, 1956): "A system of agriculture of an ethnographic unit is the customary pattern of behaviour followed by the individual members of the unit in the realm of agricultural technology, which results in typical sets of (1) land utilization in space (2) land utilization in time (3) seasonal distribution of labor and (4) seasonal distribution of nutrition and other needs."

Second, from a tropical applied agricultural research organization, ICRISAT (Krantz, 1974): "The entire complex of development, management, and allocation of resources as well as decisions and activities which, within an operational farm unit or a combination of such units, results in agricultural production, and the processing and marketing of the products."

It must be allowed that words and phrases in a succinct definition of a complex entity are forced to bear a heavy semantic load. Even so, de Schlippe's formula can be recognized as a distinctly anthropocentric definition and ICRISAT's as a distinctly economic one. Both stress the manipulation of resources by man for production, one from a sociocultural and the other from a socioeconomic viewpoint, but place less emphasis on the resource itself or on the biological and physical basis of its exploitation.

The definition of a farming system proposed here is briefer and broader, and for these reasons somewhat less specific. A farming system is defined as the *pattern of resources and processes of resource use in a farming unit.* To demonstrate that the above is not merely a form of words, some attempt must be made to describe, if not to define, what is meant by "farming unit," "pattern of resources," and "processes of resource use."

■ 1.1.3 *The farming unit.* Whatever definition of a farming system we adopt, and whatever our disciplinary mode of approach, we are concerned with specific examples or types of "farming units" or "agricultural production units." It would be pleasant, and also better English, if we could substitute for these phrases the simple word "farm," but to do so would invite confusion. The problem is that according to circumstance we may need to delineate the boundaries of our unit on geographical or economic criteria or on the basis of distinctive technology.

For the economically independent and physically contiguous farm with a defined perimeter, geographical and economic boundaries are both clear and coincident. The farm owner or tenant, with his family and perhaps some hired labor, is operating on a discrete block of land which he manages as a discrete economic unit. There is no problem.

On the other hand, what of the shifting cultivator, who one year is cropping area *A* and perhaps the next year area *B*? Both are selected by him from within a large and ill-defined area *C*, which at a later stage he may abandon for another ill-defined area *D*. In economic terms, his production unit is readily demarcated, but what is his "farm" in the physical sense? As another example, a wet rice farmer in a Southeast Asian valley with too small a crop area for subsistence may extend his operations and grow rainfed crops in the nearby hills under shifting cultivation. Geographically, he has two separate "farms" and is furthermore operating them by vastly different sets of techniques, but economically he and his family are a single unit.

Other difficulties in defining the unit arise when land is utilized communally or in sharecropping operations, or when the family generates income by off-farm employment. Such examples could be multiplied, and it is clear that the word "farm" is not of much use to us. "Farming unit" is perhaps little better, but its very lack of precision at least gives us the flexibility to apply it, in any specific instance, to a geographical unit, an economic unit, or to a unit displaying a particular technical pattern of resource use. Nevertheless, in all cases the entity as defined is a unit concerned with the output of agricultural commodities.

■ 1.2 The pattern of resources

■ 1.2.1 The individual farming unit will be characterized by a particular pattern or "mix" of resources. The components may logically be grouped into four categories:

a. *Natural resources.* The given elements of land, water, climate, and natural vegetation that are exploited by the farming unit for agricultural production.
b. *Human resources.* The human beings that live and work within the farming unit and exploit the resources at their disposal for agricultural production.
c. *Capital resources.* The goods and services created, purchased, or borrowed by the humans associated with the farming unit to facilitate their exploitation of natural resources for agricultural production.
d. *Incipient products.* The agricultural output of the farming unit at all stages in its development to the point of consumption, sale, or export from the unit.

The first three of these are, of course, the economist's classical elements of land, labor, and capital, though in subsistence farming, of major importance in tropical agricultural systems, "labor" is rather too narrow a term, as will be seen. The grouping of incipient products under "resources" also requires some justification.

■ 1.2.2 *Natural resources.* This category requires little further elaboration, for the characteristic features of general importance in agricultural production apply directly to tropical farming systems. They include:

a. The area of the farm, its topography, the degree of fragmentation of the holding, its location in respect to markets, etc.
b. Soil depth, chemical status, and physical attributes.
c. The availability of surface water and ground water.
d. The average rainfall, evaporation, radiation, and temperature pattern and its seasonal and annual variability.
e. The natural vegetation.

The significance of natural vegetation as a resource varies widely. At one extreme, for instance on a small Javanese rice farm with no herbivorous livestock, it is of no importance at all, and indeed may no longer exist, so intensive is the utilization of land for cropping. On the other hand, in pastoral systems it may constitute the whole feed resource. In shifting cultivation systems its character has a significant influence on the productivity of cropland created within it by clearing.

■ 1.2.3 *Human resources.* Several attributes are critical to the definition and characterization of the human resources of the farm unit. They include:

a. The number of persons associated with the unit that has to be fed from it in relation to the work force and its productivity, which governs the surplus available for sale, barter, or cultural obligations.
b. Capacity for work, as influenced by nutrition and health status.
c. Inclination to work, as influenced by economic status and cultural attitudes toward leisure.
d. The flexibility of the work force to adapt to seasonal variations in work demand; that is, the availability of hired labor and the degree of mutual cooperation between farming units.

Attributes (a) and (b) are of particular relevance to tropical farming systems. With respect to (a), food consumption by the farm family in the farming systems of developed regions is often low in relation to total output: for example, the average American rural worker produces enough food for fifty-five people. Furthermore, the high-technology farmer may consume little of his own output; the wheat-farmer's wife will buy bread at the local supermarket. In the tropics, however, family subsistence is the main *raison d'être* of a high proportion of farming systems. Hence the balance of workers and consumers within the farming unit is critical.

With respect to (b), capacity for work is generally taken for granted in the agricultural work force of advanced nations, but in the poorer regions of the tropics its impairment through hunger or disease is frequently a factor limiting work output.

Finally, it will have been noted that the word "cultural" appears in both (a) and (c). In the more traditional agricultural societies of the tropics, there are likely to be many activities and attitudes in relation to work that in "advanced" societies would be regarded as "noneconomic."

■ 1.2.4 *Capital resources.* For the specialist in production economics there are refined ways of categorizing capital resources, but for the student of agricultural systems, who must be at once agronomist, economist, geographer, ecologist, soil scientist, hydrologist, and social anthropologist, a simple grouping into four will suffice:

a. *Permanent resources.* That is, permanent modifications to the natural land or water resource for the purpose of agricultural production. Cleared land is perhaps the primary permanent resource (although in a shifting cultivation, cleared land is a semipermanent resource only). An established irrigation system is another.

b. *Semipermanent resources.* Capital resources that are used for agricultural production but which depreciate to the point that they require periodic replacement: for example, barns, fences, draft animals, implements, etc.

c. *Operational resources.* Consumable resources utilized in the day-to-day operations of the farming unit: for instance, fertilizer, herbicides, etc.

d. *Potential resources.* Resources that the farmer does not own but which may be commanded and which will eventually have to be repaid, whether under a formal, informal, or even an unspoken contract: e.g., credit from a bank, co-operative, or moneylender; assistance in farm work from relatives and friends, etc.

■ 1.2.5 *Incipient products.* The crops or livestock that are raised by the farming unit are a resource component until such time as they are consumed, sold, or otherwise disposed of from the farm. They represent the result of a manipulation of resources and have a value that comprises some proportion of the total economic value of the unit. Thus if the farm is bought or sold, the standing crops and the livestock on it form a specific part of the purchase price.

■ 1.2.6 The definition of a farming system from ICRISAT extends to the processing and marketing of the products of the farming unit. Processing of products is certainly a part of the activities of the farm unit if it takes place within its human and geographical boundaries. The marketing element is not strictly part of the *farming* system, though the demands of the market will necessarily influence the type of crop produced.

■ 1.3 The processes of resource use

■ 1.3.1 The phrase "processes of resource use" in our definition of a farming system denotes the physical and biological processes or subsystems that underlie farming operations—the manipulation of resources for production. Since the mode of operation of agricultural practices can be described in terms of the management and exploitation of energy, water and soil nutrients, we can group the processes of resource use into: (a) energetic processes, (b) hydrological processes, and (c) biogeochemical processes.

Each set of processes can be evaluated in terms of inputs, outputs, storages, and transformations.

■ 1.3.2 *Energetic processes.* There are six main sources of energy input into a farming system: solar, human, draft animal, food, fuel, and indirect energy. Solar energy is a product of the natural resource, human energy of the human resource. Food and fuel energy may be an input if they are purchased; indirect energy is that sequestered in purchased manufactured goods such as implements. In high-technology farming systems, the magnitude of fuel and indirect energy inputs into farming operations has been brought sharply into focus in recent years with a realization of the finite limits and precarious supply of the world's fossil fuels, but it is of far less importance in tropical agriculture because of the limited degree of mechanization and limited use of agricultural chemicals. Conversely, human and draft animal energy assume a far greater significance in tropical farming systems.

The major energy outputs, apart from the losses and inefficiencies in energetic transformations that are dissipated as heat and which contribute to the running-down of the universe, are of course the energy in crops and livestock consumed or sold.

Energetic transformations within the farming system are numerous. Solar energy is fixed as plant tissue energy through photosynthesis: plant energy is converted to human and animal tissue energy by consumption, or it may be more directly transformed into human or draft animal work energy. Animal tissue energy is utilized by humans when they consume stock or stock products; fuel energy is converted to power by machinery, etc. Animal energy may be utilized as fuel energy when dung is burned.

The concept of energy storage within the farming unit is less readily grasped than, say, the storage of water or nutrients, but it can be equated with the energy sequestered in the vegetation, crops, and livestock present at any one time within the unit. This store of energy may be realized by the sale of products, by human or animal consumption, or, as in the case of shifting cultivation systems where vegetation is burned, for the maintenance of soil fertility.

■ 1.3.3 *Hydrological processes.* Since water, unlike energy or nutrients, does not undergo transformation in farming operations other than into water vapor, the hydrological process can be viewed solely as inputs, outputs, and storages, the components of the hydrological cycle. The inputs into the farming system are precipitation, run-on, and irrigation water; the outputs evaporation, runoff, and drainage beyond the effective root zone of plants. To outputs, we can add water consumed by the farm humans and livestock. In magnitude this is small, but the availability of drinking water in space and time has a marked effect, for example, on the character of pastoral systems.

The storage component in the farm hydrological system includes water stored in the soil profile for direct use by crops and vegetation, groundwater that may be drawn up for use by humans, livestock, or crops, and water in constructed storages such as farm ponds.

■ 1.3.4 *Biogeochemical processes.* The major biogeochemical inputs into the farming system are the available nutrients released from the soil nutrient store, the fixation of atmospheric nitrogen by legumes, nonsymbiotic nitrogen fixation (of particular importance in rice-growing), nutrients in rainfall and run-on water, fertilizer, and nutrients in purchased human food, stock feed, or animal manure. The purchase of stock feed or manure is uncommon in tropical farming systems except of the most intensive type.

The important outputs include nutrients in crop and livestock or livestock products consumed or leaving the farm. Other outputs or losses inherent in farming operations are associated with leaching beyond the root zone, denitrification and volatilization of nitrogen, losses of nitrogen and sulfur to the atmosphere when vegetation is burned, nutrients lost in soil erosion caused by runoff or wind, and nutrients in human or livestock excreta that are for some reason lost from the farm and not recycled.

The biogeochemical storage component comprises fertilizer stored and manure accumulated, together with the sum of nutrients in the soil root zone, in the standing crop and vegetation, and in livestock.

As with energy, biogeochemical transformations are many and varied. Soil nutrients undergo transformation from the unavailable to the available form and vice versa; available soil nutrients are taken up by crops and vegetation. Crop and vegetation nutrients are cycled through humans and livestock by consumption of plant products and are returned to the soil-plant system in dung and urine, or reach the same end-point by a more complex route when stock or stock products are consumed by the farm family. In farming systems involving the burning of vegetation, the vegetation nutrient store is periodically returned to the soil for crop uptake or is partially lost to the atmosphere.

■ 1.3.5 *The socioeconomic context.* The subsuming of the processes of resource use under physical and biological categories only may be questioned on the general grounds that farming is an economic activity and involves economic processes and subsystems. However, it is more proper to regard economic processes as abstractions, within a common framework of terms and units, of real-life physical and biological processes. On the other hand, we can recognize sets of economic and social *conditions* of critical significance in the biological and physical exploitation of resources for agricultural production. Within the framework of this book, an appropriate designation of these might be the *socioeconomic context* of farming systems.

■ 1.4 Farming System Typology

■ 1.4.1 At the beginning of this chapter it was suggested that the first serious study of farming systems involved a classification of types by geographers. The problems of categorization and of the criteria to be used in distinguishing classes have been a continuing source of controversy.

Whittlesey (1936), for example, put forward the following five distinguishing criteria in his now classic treatment of the topic:

a. The crop and livestock association.
b. The methods used to grow the crops and produce the stock.
c. The intensity of application to the land of labor, capital, and organization, and the outturn of the product which results.
d. The disposal of the products for consumption.
e. The ensemble of structures used to house and facilitate farming operations.

As can be seen, the criteria are varied in that some are based on resources, some on processes, and some on a combination of both.

The debate has continued, to the point where in 1964 the International Geographical Union set up a Commission on Agriculture Typology in an attempt to formulate common taxonomic criteria and classes. A summary review of progress has been made by Kostrowicki (1977).

As an instance of recent broad-scale farming system taxonomy, Grigg (1974) groups the agricultural systems of the world under the following headings (for which he disclaims comprehensiveness): (a) shifting agriculture, (b) wet rice cultivation, (c) pastoral nomadism, (d) Mediterranean agriculture, (e) temperate mixed farming, (f) dairying, (g) the plantation system, (h) ranching, and (i) large-scale grain production.

It is clear that the classes are designated by a range of criteria: products, climatic type, and economic and managerial structure. Perhaps this is inevitable when one is pursuing a will-o'-the-wisp: that is, a global general-purpose typology. As Grigg says regretfully, "pragmatism must take precedence over principle."

■ 1.4.2 Narrowing our focus to tropical farming systems only, the most serviceable classification is that of Ruthenberg (1971), who recognizes seven main types: (a) shifting cultivation systems, (b) semipermanent rainfed cultivation systems, (c) permanent rainfed cultivation systems, (d) arable irrigation systems, (e) perennial crop systems, (f) grazing systems, (g) systems with regulated ley farming.

The term "permanent cultivation" denotes the continuous or near-continuous annual cultivation of a given land area, as distinct from "shifting cultivation," which signifies periodic cropping only of a specific land area. "Semipermanent" describes rainfed cropping systems where the frequency with which any specific land area is cropped is intermediate between that of shifting cultivation and permanent cropping. System (g), regulated ley farming, that is, the organized alternation of a phase of arable cropping and a phase of improved sown pasture, is of minor importance in the tropics.

In this book, Ruthenberg's categories are retained, as are the quantitative distinctions in cultivation frequency between (a), (b), and (c) (see section 1.4.3). Since we are confining our attention to annual cropping systems, only the first four classes are relevant. However, the word "permanent" has connotations beyond that of frequency of cropping, and "intensive" is here preferred. The classes adopted are, therefore: (a) shifting cultivation systems, (b) semi-intensive rainfed systems, (c) intensive rainfed systems, (d) irrigated systems.

■ 1.4.3 *Cultivation frequency.* In detailed classifications of tropical rainfed annual cropping systems, categories are distinguished by reference to the frequency with which a given land area is cropped. The criterion scarcely applies to irrigated systems since if irrigation water is available the land is normally cropped every year.

Quite elaborate typologies have been created (e.g., Allan, 1965) based on the proportion of cropland to "fallow" land, the term normally used to describe the phase between successive cropping periods when native or adventive species recolonize the abandoned cropfield. Ruthenberg (1971) uses the characteristic *R*, representing the length of the cropping phase in years divided by the total length of the cultivation cycle, crop years plus fallow years, expressed as a percentage. Thus if in a shifting cultivation system the land is cropped for one year and fallowed for nine years, the *R* value is 10. Ruthenberg sets the boundary between shifting cultivation and his semipermanent cropping at $R = 30$, and that between semipermanent and permanent cropping at $R = 70$. These limits are retained in the present volume to distinguish shifting, semi-intensive, and intensive rainfed systems, but the term "cultivation frequency" is used in place of the characteristic *R*.

Cultivation frequency when applied to the less intensive annual cropping systems denotes the proportion of years for which the land is cropped, without reference to the number of cropping periods within the year. In some of the more complex shifting cultivation systems in regions with little or no dry season it would be impossible to designate the number of cropping periods, since planting and harvesting are virtually continuous. However, for intensive systems where land is cropped every year and cropping cycles are more orderly, some index is needed to designate the intensity of cropping within the year. "Multiple cropping index" is one term in use, but in this book the simpler phrase "cropping index" is preferred. Some difficulties arise in respect of the degree to which crop cycles overlap in time, but for the most part the designation is serviceable.

■ 1.5 Type and process: a two-way table

It is clear that we can analyze and describe farming systems by ordering our information in two dimensions: on the basis of types of farming unit, or on the basis of biological and physical processes. In this book the two approaches are combined: the warp being a three-part classification of types (semi-intensive and intensive rainfed cropping systems are treated together), the weft being the three-part classification of processes: energetic, hydrological, and biogeochemical. For reasons partly pedagogic and partly related to the amount of data available, equal stress is not placed on all nine of the resulting compartments. Figure 1.1 shows where the main emphases are laid.

	Biogeochemical processes	Energetic processes	Hydrological processes
Shifting cultivation systems	■		
More intensive rainfed systems		■	
Irrigated systems			■

Figure 1.1. Types and processes of farming systems.

■ 1.6 CLIMATES AND SOILS OF THE TROPICS

To conclude this introductory chapter, and to provide some broad quantitative background on climates and soils in the tropics, Tables 1.1 and 1.2 are presented. Table 1.1 shows land areas within five major climatic zones in the tropics, distinguished on the basis of the number of wet months per year. The aridity limits to cropping in the tropics lie within zone (d), which has 2 to 4.5 wet months; hence at the minimum the area of land with adequate rainfall for cropping is 3.6×10^9 ha. Table

TABLE 1.1
Land areas of major climatic regions of the tropics
(ha $\times$ 10^6)

Rainfall regions	Asia and Pacific	Australia	Africa	America		Total
				S	N	
a. Wet 9.5–12 m	348	—	197	601	45	1,191
b. Wet 7–9.5 m	275	9	499	576	51	1,410
c. Wet 4.5–7 m	148	53	645	98	76	1,020
d. Wet 2–4.5 m	87	114	487	69	15	772
e. Wet $<$ 2 m	131	98	305	19	7	560
Total	989	274	2,133	1,363	194	4,953

SOURCE: Engelstad and Russel, 1975, adapted from the Report by the President's Science Advisory Committee on the World Food Problem.

TABLE 1.2
Distribution of major soil types by climatic regions in the tropics (ha $\times 10^6$)

	Rainfall regions (see Table 1.1)					
	a	b	c	d	e	Total
1. Light-colored soils, base-rich; Aridisols	5	7	98	378	211	699
2. Dark-colored soils, base-rich; Mollisols and Vertisols	23	56	119	93	1	292
3. Moderately weathered and leached soils; Alfisols and Andepts-Inceptisols	4	34	90	75	6	209
4. Highly weathered and leached soils; Oxisols and Ultisols	931	1,084	474	49	1	2,539
5. Shallow soils and dry sands; Entisols	81	105	170	153	336	845
6. Alluvial soils; Entisols	146	124	71	23	5	369
Total	1,190	1,410	1,022	771	560	4,953

SOURCE: Engelstad and Russel, 1975, adapted from the Report by the President's Science Advisory Committee on the World Food Problem.

1.2 gives the distribution of the main tropical soil types within these five climatic zones. Excluding type (5) as potentially nonarable reduces the area within climatic zones (a), (b), and (c) by about 0.35×10^9 ha. Thus on the most approximate basis, without taking into account altitude, topography, drainage, and land alienated for other purposes, the area within the tropics with a climate and soil potentially suitable for cropping is of the order of 3×10^9 ha.

REFERENCES

Allan, W. 1965. *The African husbandman.* Edinburgh and London: Oliver and Boyd.

Boserup, E. 1965. *The conditions of agricultural growth.* Chicago: Aldine.

Brookfield, H. C., and Brown, P. 1963. *Struggle for land.* Melbourne: Oxford University Press.

Clark, C., and Haswell, M. 1970. *The economics of subsistence agriculture.* 4th ed. London: Macmillan.

Dent, J. B., and Anderson, J. R., eds. 1971. *Systems analysis in agricultural management.* Sydney and New York: Wiley, Australia.

Duckham, A. N., and Masefield, G. B. 1971. *Farming systems of the world.* London: Chatto and Windus.

Engelstad, O. P., and Russel, D. A. 1975. Fertilizers for use under tropical conditions. *Advances in Agronomy* 27:175.

Grigg, D. B. 1974. *The agricultural systems of the world. An evolutionary approach*. London: Cambridge University Press.

Haswell, M. R. 1973. *Tropical farming economics*. London: Longman.

Kostrowicki, J. 1977. Agricultural typology concept and method. *Agricultural Systems* 2:33.

Krantz, B. A., and Associates. 1974. *Cropping patterns for increasing and stabilizing agricultural production in the semi-arid tropics*. Hyderabad, India: International Workshop on Farming Systems. ICRISAT.

Nye, P. H., and Greenland, D. J. 1960. *The soil under shifting cultivation*. Harpenden, England: Commonwealth Bureau of Soils, Technical Communication no. 51.

Ruthenberg, H. 1971. *Farming systems in the tropics*. Oxford: Clarendon Press.

de Schlippe, P. 1956. *Shifting cultivation in Africa. The Zande system of agriculture*. London: Routledge and Kegan Paul.

Spedding, C. R. W. 1975. *The biology of agricultural systems*. London and New York: Academic Press.

Whittlesey, D. 1936. Major agricultural regions of the earth. *Annals of the Association of American Geographers* 26:199.

Chapter 2

General Hydrological Background

■ 2.1 Tropical rainfall regimes

■ 2.1.1 There is a variety of global climatic classifications to choose from in defining what is meant by the terms "tropical," "semi-arid," "subhumid," and "humid," which, however, need not worry us. (Trewartha, 1968, gives a straightforward classification and characterization of the major tropical climates.) Essentially we are concerned here with areas of the earth where the mean temperature of the coldest month at marine locations exceeds 17°C and where mean annual rainfall ranges from over 3,000 mm down to about 400 to 500 mm, though there are tropical regions where crops are grown under a mean annual rainfall of 350 mm.

Whatever the subdivisions used by climatologists to categorize tropical rainfall regimes—Trewartha's *Ar* and *Aw*, for example—the first important point is that the range of precipitation patterns encountered is continuous. As one moves away from the wet tropical equatorial zones to higher latitudes, the trend toward a lower mean annual rainfall is normally accompanied by a lengthening period of dry weather in winter as the intertropical convergence zone, with a slight time lag, "follows the sun," giving rise to the summer-rainfall regimes described as monsoon or savannah climates. The second important point is that, with decreasing mean annual rainfall, variability in total rainfall and in the time of onset and cessation of the rainy period increases, and the probability of occurrence of drought spells within the rainy period also increases.

■ 2.1.2 Within this general pattern, the main variants in climate type that are important in relation to the character of tropical farming systems are these:

a. twin-peak rainfall regimes commonly found in regions transitional between the true wet tropics and the summer-rainfall tropics (e.g., Colombo, Sri Lanka).
b. those summer-rainfall regions that have an appreciable winter rainfall component associated with pressure systems distinct from the summer monsoon (e.g., N. India).
c. high-altitude locations, where the lower temperatures appreciably reduce the evapotranspiration component of the water balance (e.g., the tropical Andes).

■ 2.2 The average annual water balance

■ 2.2.1 As a basis for appreciating the seasonal hydrological pattern in tropical farming systems, we must first briefly remind ourselves of the concepts of the water balance, which over a specific period may be expressed as:

$$\triangle M = P - (O + U + E_a)$$

where $\triangle M$ = net change in soil water in the crop root zone during the period specified;
P = precipitation, which in the tropics is of course almost wholly rainfall;
O = runoff;
U = drainage below the root zone;
E_a = actual crop evapotranspiration.

When the water balance concept is used in agronomy and crop physiology, interest is centered on the water that is used by the crop; that is, on the terms P and E_a. When P exceeds E_a and the root zone is fully charged, resulting in runoff or drainage, the agronomist or crop physiologist is rarely concerned with the magnitude of these latter components. However, in the study of farming systems they are as important as the other terms. The size of the drainage term has a marked influence on the leaching of soluble nutrients; the incidence of runoff, in relation to slope and topsoil condition, governs the topographical siting of cropfields and their liability to soil erosion.

■ 2.2.2 To grasp the general span of seasonal hydrological patterns encountered in tropical crop farming, let us first consider a brief series of type locations. Figure 2.1 illustrates the average annual pattern of rainfall and evaporation at five sites in South and Southeast Asia. It cannot be stressed too strongly that we are here considering long-term averages. The probability that the hydrological pattern in any one year

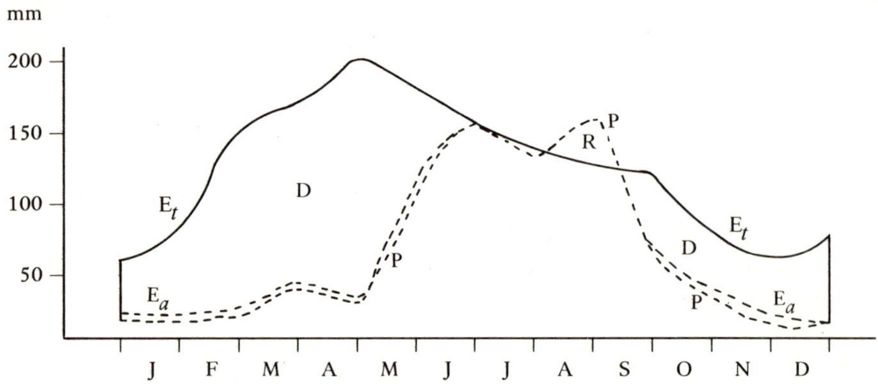

Figure 2.1a. Hyderabad. Average seasonal water balance (Lockwood, 1974). D—moisture deficit; R—moisture recharge; P—precipitation; E_t—potential evapotranspiration; E_a—actual evapotranspiration.

will approximate to the average is always low and decreases with decreasing total annual rainfall.

At Hyderabad, exemplifying the semi-arid tropics (mean annual rainfall 772 mm), rainfall exceeds potential evaporation on average for only a brief period toward the end of the summer wet season. Recharge of the profile during this period does not, on average, lead to any appreciable surplus for runoff or drainage, though in any one season there may be substantial runoff or drainage during individual rain periods. In such a region, farming systems are based on a single short-season monsoon or *kharif* crop such as millet or sorghum, though the favorable water retention characteristics of some soil types in this area permit crops to be grown well into the dry season.

Cuddalore and Satkhira represent summer-rainfall regions of increasing amount and duration of precipitation. At Cuddalore, on average, the profile is fully recharged about halfway through the period during which P is greater than E_t; at Satkhira this occurs about one-third of the way through the period. In both instances, therefore, there is a substantial water surplus which is not utilized for crop growth unless it is impounded for irrigation either at the site or further down the catch-

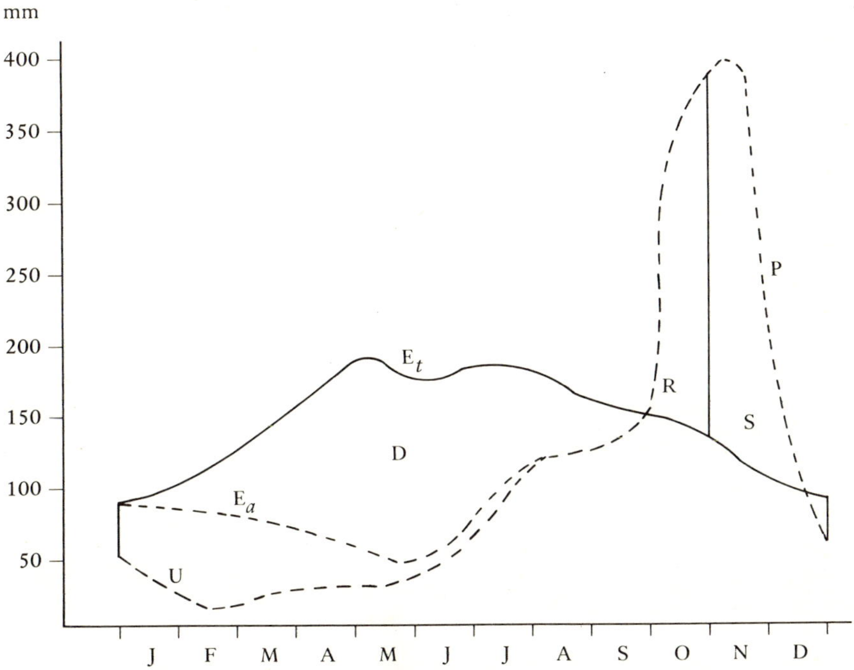

Figure 2.1b. Cuddalore. Average seasonal water balance (Lockwood, 1974). D—moisture deficit; U—moisture utilization; R—moisture recharge; S—water surplus; P—precipitation; E_t—potential evapotranspiration; E_a—actual evapotranspiration.

ment. Such hydrological regimes permit cropping systems based on longer-season crops or on double or relay cropping.

Colombo is a typical example of a classical twin-peak rainfall regime, where the pattern of soil water deficit and utilization during the dry phase and recharge and surplus during the wet phase occurs, in the average situation, twice in a year. In this instance the deficit periods are relatively brief, and with continuously favorable temperatures year-round cropping is possible. The prolonged periods of surplus enforce in low-lying areas cropping systems based on a flood-tolerant crop, that is, wet rice.

Finally, Singapore represents the rainfall and evaporation pattern of the true wet tropics, where *P* on average exceeds E_t all the year round. The island of Singapore can scarcely be described as a typical agricultural location, but the regime is representative of equatorial regions where perennial moisture-loving crops such as rubber and oil palm are

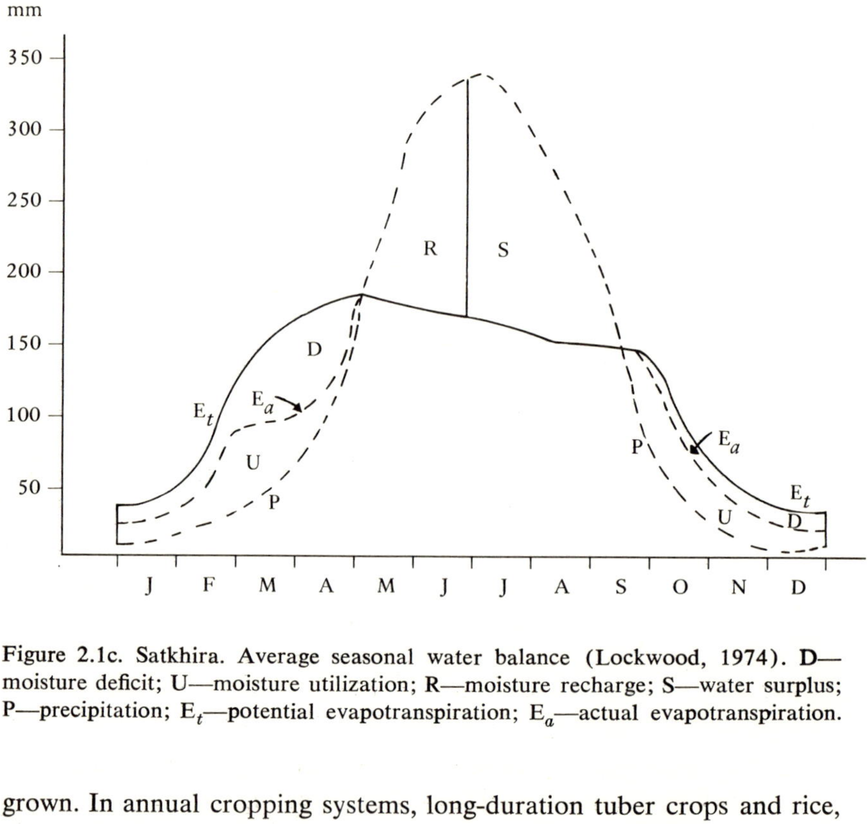

Figure 2.1c. Satkhira. Average seasonal water balance (Lockwood, 1974). D—moisture deficit; U—moisture utilization; R—moisture recharge; S—water surplus; P—precipitation; E_t—potential evapotranspiration; E_a—actual evapotranspiration.

grown. In annual cropping systems, long-duration tuber crops and rice, on a virtually nonseasonal basis, are typical components.

■ 2.3 Average length of the growing season

■ 2.3.1 In consideration of Figure 2.1, reference was made to the length of the period during which *P* exceeded E_t. This can be used as a measure of the crop growing season, but it is inadequate to charac-

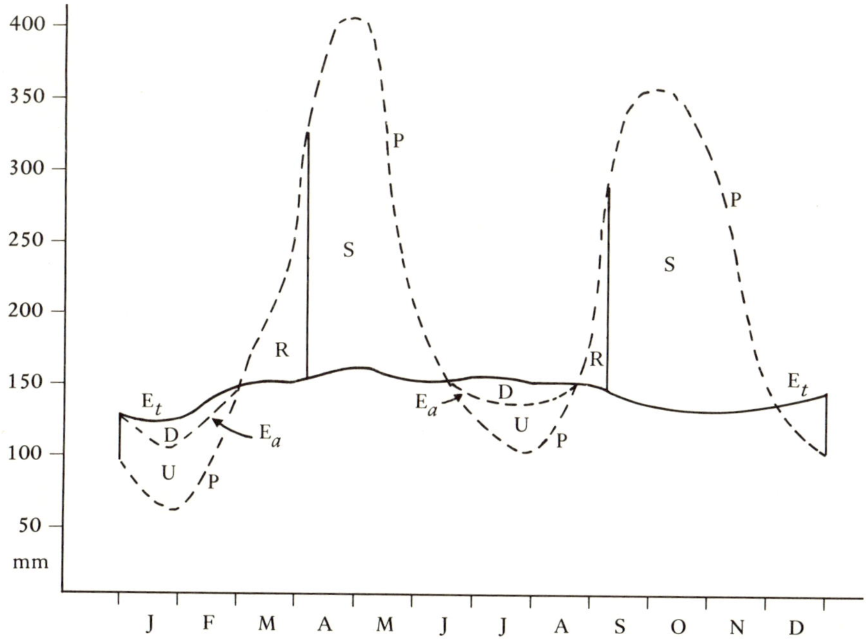

Figure 2.1d. Colombo. Average seasonal water balance (Lockwood, 1974). D—moisture deficit; U—moisture utilization; R—moisture recharge; S—water surplus; P—precipitation; E_t—potential evapotranspiration; E_a—actual evapotranspiration.

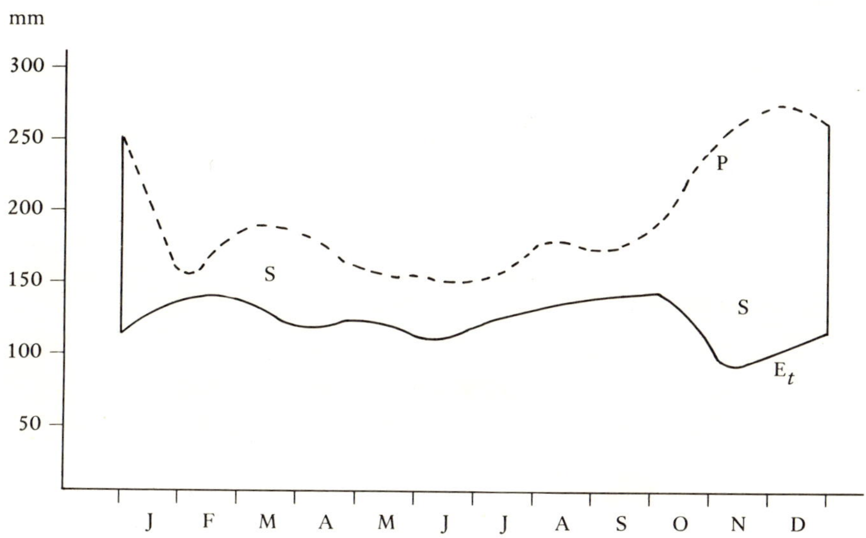

Figure 2.1e. Singapore. Average seasonal water balance (Lockwood, 1974). S—water surplus; P—precipitation; E_t—potential evapotranspiration.

terize the average pattern of water availability for crop farming operations. A more detailed approach is that of Cochemé and Franquin (1967), which takes into account the land preparation phase preparatory to cropping and the residual soil water available for crop maturation after the rains have ended.

Given a climate with a well-defined wet and dry season—the analysis of Cochemé and Franquin is based on West African data—the growing season may be divided into five periods, defined by P and E_t:

a. *Preparatory period*: from when $P = E_t/10$ to when $P = E_t/2$. This is regarded as a phase during which land preparation can proceed.

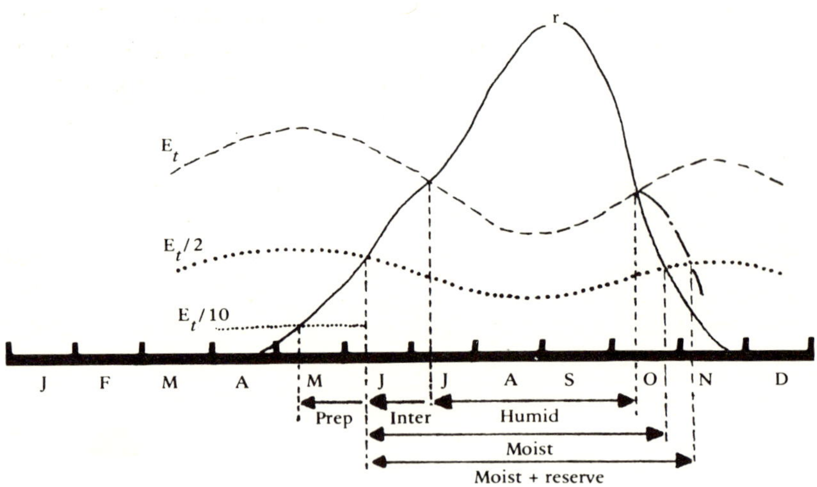

Figure 2.2. Significant phases of the water balance in a single-peak summer rainfall climate (Lockwood, 1974; after Cochemé, 1968). E_t—potential evapotranspiration; r—rainfall.

b. *First intermediate period*: from when $P = E_t/2$ to when $P = E_t$, during which the crop may be sown or planted.

c. *Humid period*: when P exceeds E_t, the dependable period for crop growth and development.

d. *Second intermediate period*: when, as the rains are declining, P falls to between E_t and $E_t/2$. Such conditions, coupled with a charged soil profile, are adequate for continued crop growth.

e. *Reserve period*: from when $P = E_t/2$ to when $P = E_t/10$. In this phase rainfall effectiveness is low and continued crop growth and maturation are largely dependent on the available soil water reserve.

■ 2.3.2 Figure 2.2 shows a typical pattern. The *effective crop growing season* is defined as the period from when P exceeds $E_t/2$ until the soil water reserve, assuming a standard 100 mm of available water in the crop root zone, is exhausted; that is, periods (*b*) through (*e*).

The average duration of the individual phases for a range of stations in West Africa, ranked in order of mean annual rainfall, is shown in Figure 2.3. The two intermediate periods are of relatively constant duration—25 days for the first and 15 days for the second—but the humid period increases from zero to 160 days from the driest to the wettest station. For the latter, the effective crop growing season as defined above is 200 days, long enough under rainfed conditions for a main crop and a short-season second crop, or for a long-season tuber crop such as yam.

An illustration of the effect of increasing length of growing season is given in Table 2.1, where a north-south transect in West Africa is grouped into four regions on the basis of increasing rainfall. The main and secondary crops grown in the region are listed. Note the shift in cereal crop from millet to sorghum to corn with increasing length of growing season (moving south), and the changes in maturity type of the crops.

■ 2.3.3 One of the sources of imprecision in estimates of the length of growing season based on climatic data alone is the assumption made of the available soil water reserve. In the absence of local soil physical data, it is usual to postulate a standard figure of 100 mm. However, differences in water characteristics between soils at a given location may have a marked effect on the length of the effective crop growing season and hence on the character of cropping systems, as at Hyderabad (see section 2.2.2).

For example, McCown (1973) examined the effect of available water capacity on length of growing season on three soils at Townsville, North Australia (about 750 mm rainfall). With experimental data obtained at four sites over four years, he generated a model that simulated length of growing season over 60 years of meteorological records for soils of 75, 150, and 180 mm available water. The results indicated little

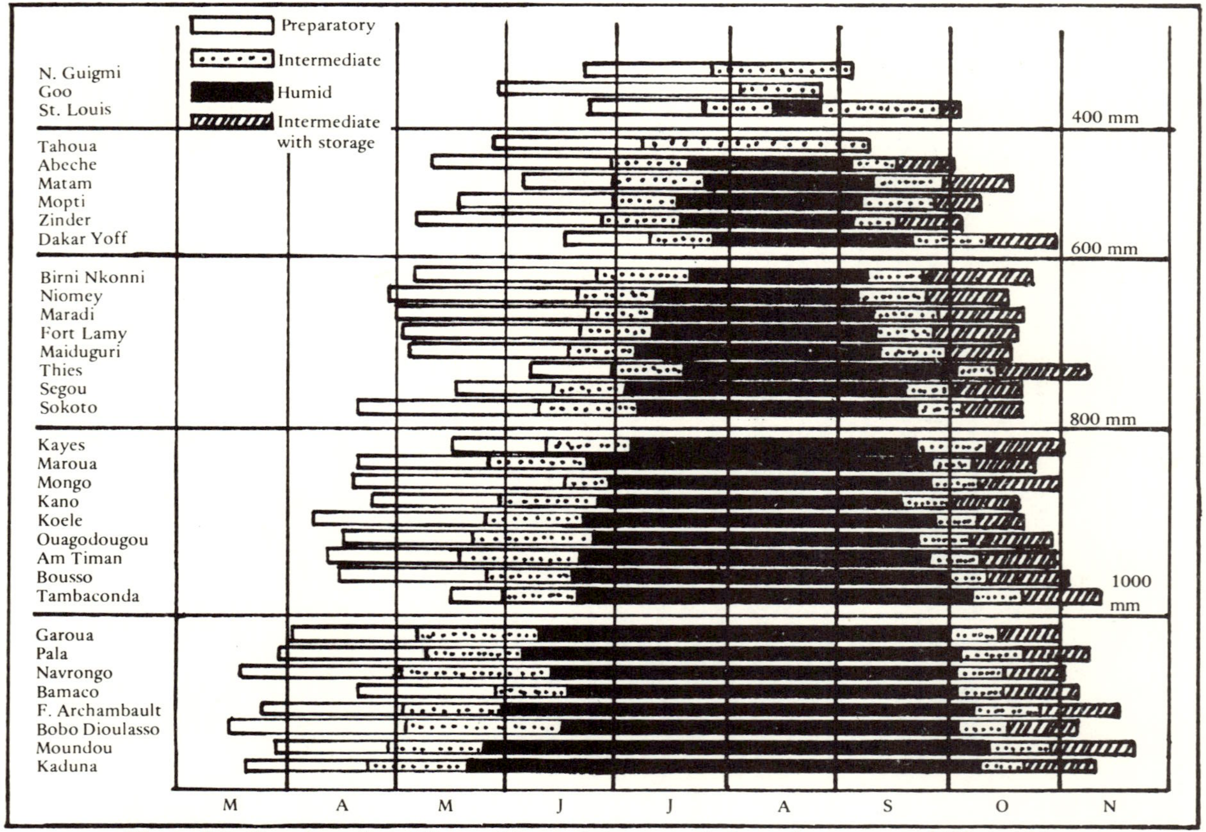

Figure 2.3. Duration of significant phases of the water balance in a range of West African rainfall regimes (Cochemé and Franquin, 1967).

effect in very wet and very dry years, but for the central 45 percent of the years, when the growing season varied from 70 to 110 days on the 75 mm soil, it averaged 14 days longer on the 150 mm soil and 20 days longer on the 180 mm soil.

■ 2.4 The stochastic element in the hydrological pattern

■ 2.4.1 Sections 2.2 and 2.3 were concerned only with the *average* rainfall and evaporation pattern and with the *average* duration of phases of the growing season. The variability between years in these

TABLE 2.1
Principal crops of rainfed farming systems in four rainfall zones of West Africa

				Crops	
Region	MAR*	EAR	LGS	Main	Secondary
A	350	250	60	Cowpea SS millet	Cowpea
B	550	475	100	MS millet SS, MS sorghum SS, MS peanuts	SS millet SS sorghum Cowpea
C	800	600	120	MS, LS millet MS, LS sorghum Cotton	SS millet SS sorghum SS corn Cowpea
D	1,000	750	145	LS sorghum LS corn LS millet LS peanuts Cotton	SS millet SS sorghum SS corn Cowpea

Source: After Cochemé and Franquin (1967).
*MAR = mean annual rainfall (mm); EAR = effective annual rainfall (mm); LGS = length of growing season (days); SS = short-season; MS = mid-season; and LS = long-season.

parameters, and the fact that this variability tends to increase with decreasing mean annual rainfall, has already been emphasized. The concepts illustrated in Figures 2.1 and 2.2, though useful for broad-scale resource evaluation, are of limited value in relation to the operations of cropping. Essentially the seasonal pattern of crop farming in any given location, insofar as it is governed by the availability of water, represents an adaptation to the *probability of occurrence* of hydrological events and situations at different times of the year.

■ 2.4.2 Given the appropriate long-term data series, or estimates from such data, we may for example compute:

a. The probability of occurrence of dry spells of any defined duration and intensity at any defined period during the wet season;
b. The probability of occurrence of defined quantities of runoff at any defined period during the wet season;
c. The probability of adequate soil water for land preparation or for sowing at any defined date at the start of the wet season;
d. The probability of exhaustion of the soil water reserve at any defined date at the end of the wet season;
e. The probability of occurrence of a defined length of growing season, depending on our definition of it (that of Cochemé and Franquin, 1967, given in section 2.3.2, is only one example).

■ 2.5 Crop water balance models

It will be appreciated that hydrological evaluation of the growing season at a particular site based on rainfall and evaporation data alone, with measurements or estimates of soil water-holding capacity, is deficient in that it does not take into account the water requirement of the crops or crop types that are grown or could be grown at the site. To estimate changes in water available for crop growth in the simulated situation of a growing crop, crop water balance models are employed. To obtain the E_a term of the water balance equation involves estimates of E_t (in relation to E_o) by the defined crop or crop type at different stages in its development—related largely to canopy cover (Ritchie and Burnett, 1971)—and of the ratio of E_a to E_t, as influenced by the evaporative demand of the atmosphere and available soil water in the root zone (Ritchie, 1973).

This book is not the place to discuss such models in detail. Essentially they are designed to estimate for a given environment, by daily or weekly water budget accounting, the changes in available soil water during the potential cropping period, from which predictions can be made concerning the adaptation of rainfed crop species and cultivars, the irrigation requirements of crops, and the appropriateness of agronomic practices such as modification of planting time. An application of such methods to cropping systems in the tropics (or, more precisely, on the tropic) is illlustrated by the work of Fitzpatrick and Nix (1969), who simulated weekly changes in soil water from limited climatic data in alternating crop-fallow systems of Central Queensland, the crops concerned being cotton, grain sorghum, and wheat. Models that utilize limited climatic data are of particular value in the tropics, since meteorological and soil physical records for the locations to which the predictions are intended to apply are frequently minimal. The principles of water balance models and their application to crop production strategy in the tropics are discussed by Basinski (1974).

■ 2.6 INTERACTIONS OF WATER REGIME AND TEMPERATURE

■ 2.6.1 In this chapter we are concerned with water use in cropping systems, and hitherto the relations between climatic regime and cropping pattern have been discussed solely in terms of the seasonal availability of water. There are, however, features of the interaction between water availability and temperature in determining the general character of cropping systems that may be appropriately dealt with at this stage.

Within the lowland wet tropics near the equator, seasonal temperature has little influence on the pattern of cropping, since it is uniformly high year-round. At higher latitudes, in predominantly summer-rainfall climates, there are distinct summer and winter temperature regimes. In lowland locations at these latitudes, it is reasonably safe to generalize that for rainfed cropping, temperature is still of minor importance, since temperatures are high when the rains begin and are still high when they end. In other words, the growing period is not limited by temperature; the character of the crop pattern is determined by the length of the period of available water. However, in lowland summer-rainfall climates where cropping is possible in the cool winter season because of good soil water storage, a winter rainfall component, or irrigation, the winter temperature regime has a strong influence on the type of crop grown.

Thus in lowland summer-rainfall regions with only a moderate variation in temperature between summer and winter, e.g., Central and North Thailand, it is possible with irrigation to grow in winter what are regarded as "summer" crops: rice, soybeans, peanuts, etc.; the same crops that under appropriate drainage conditions are grown in summer. On the other hand, in more continental lowland summer-rainfall regions with colder winters, e.g., the valleys of North India and Pakistan, only temperate crops can be grown; a typical cropping system might be summer rice–winter wheat. A detailed account of cropping patterns in relation to rainfall and temperature in India, a predominantly lowland summer-rainfall region, is given in a publication of the Indian Council of Agricultural Research (ICAR), 1970.

It is worth noting here that throughout this book the term "summer" is applied to the season of maximum sun angle. On the Indian subcontinent "summer" describes the hot period in late dry season, before the monsoon or *kharif* season. In the terminology of this book, the monsoon or *kharif* season is summer, the *rabi* season is winter or early dry season, and the "summer" the late dry season.

■ 2.6.2 In the tropics the altitudinal limits of cropping, operating through temperature, are of course substantially higher than in temperate zones: in East Africa, Papua–New Guinea, and the tropical Andes it is over 2,500 m. With reduced crop evaporation and some cloud condensation, the effective crop growing season for any given duration of rainfall is extended. On the other hand, crop maturity may be delayed by low temperature, as in Andean corn cultivars. Adaptation to low temperature may be through the evolution of high-altitude ecotypes of tropical crops (for example, sweet potatoes in New Guinea) or by growing temperate crops (as in Ethiopia, where barley is grown up to 3,000 m [Westphal, 1975]).

In highland locations water regime and altitudinal temperature effects may interact in a complex manner. A good example is in the hills of North Thailand, where non-Thai ethnic groups practice shifting cultivation, growing opium as a cash crop, corn for their pigs and poultry, and upland rice for their own sustenance. Rainfall is at least 1,500 mm over six summer months, and with good soil water characteristics and low evaporation in early winter, the effective crop growing season at high altitudes is about 7 to 8 months. Rice is a summer crop, but with the cultivars available is limited by temperature to below 1,000 m; corn is also grown in summer but will tolerate the lower temperatures above 1,000 m; opium is a warm-temperate crop adapted to the fairly cool conditions of late summer–early winter above 1,000 m.

The resolution of these temperature tolerances and water availability is that above 1,000 m the farmers grow two crops a year—corn sown in early summer, opium sown after corn in late summer and maturing largely on stored soil water in early winter—while below 1,000 m they grow a single crop of upland rice. The villages are often located near the 1,000 m level so that both sectors of the farming system may be operated with the minimum of hill-climbing.

■ 2.7 Runoff and soil erosion

■ 2.7.1 It has already been stressed that the magnitude and seasonal incidence of the runoff component of the water balance has a profound influence on the character of farming systems. This influence is twofold. It may be direct, in that the surface water accumulation and its subsequent retention or disposal govern the type of crop grown, the topographical layout of cropfields, and their surface geometry. Or it may

TABLE 2.2
Runoff and soil erosion from small cropped catchments. Hyderabad, India

Area of catchment (ha)	Average slope (%)	Rainfall (mm)	No. of runoff storms	Runoff (% of rainfall)	Soil erosion (t ha^{-1})
3.5	0.6	731	21	6.2	3.0
4.1	1–1.5	735	15	1.5	2.2
4.4	1.1	739	?	9.6	9.6
2.1	0.8	734	16	6.4	2.9
9.1	1–2.0	739	22	8.0	3.9
6.5	1.7	757	?	11.9	11.3
8.1	1.8	755	26	16.1	13.3

Source: Kampen (1974).

be indirect, acting through the erosive effect of runoff on the topsoil, which in turn governs the adaptations of the farming system for erosion control or affects the long-term stability of the system if erosion is not controlled. These influences are considered in later chapters.

Figure 2.1 indicates the relative importance of the surplus component of the average seasonal water balance in a range of rainfall regimes. In section 2.2.2 it was pointed out that even in semi-arid climates where on average the seasonal surplus is small, significant runoff can occur following individual periods of intensive rain. This is illustrated in Table 2.2 by data from seven small cropped catchments, varying in size from 2 to 9 ha, at ICRISAT, Hyderabad (Kampen, 1974). Although total annual rainfall, which is concentrated in five summer months, averaged only about 750 mm, from 15 to 26 storms generated runoff. The *runoff coefficient*, the total annual runoff as a percentage of total annual rainfall, varied from 1.5 to 16.1 percent and the resulting soil erosion from 2 to 13 t ha^{-1}.

In any series of sites of comparable soil, slope, and vegetative cover, an increase in annual amount and duration of rainfall will be associated with an increase in the number of storms that generate runoff and may also be associated with an increase in individual storm intensity (though in Africa, at least, variation in storm intensity between regions does not appear to be closely associated with mean annual rainfall; Hudson, 1971). Furthermore, in a high-rainfall regime E_a is reduced because of the low evaporative demand of the atmosphere. Hence as annual rainfall increases, the runoff coefficient increases; *a fortiori*, the absolute amount of runoff becomes very large. Thus at Hong Kong, with an annual mean rainfall of 2,128 mm, the runoff coefficient averages 43 percent, representing over 900 mm of water (Lockwood, 1974), which is more than the annual rainfall of Hyderabad.

■ 2.7.2 In the tropics, rainfall is of a higher general intensity than in temperate regions. This is illustrated in Figure 2.4. As a consequence, tropical rainfall has a higher *erosivity*, or potential capacity to cause erosion. As a common rule, rain becomes erosive when its intensity exceeds 25 mm hr^{-1}. Figure 2.4 indicates that whereas only about 5 percent of temperate rainfall is by this criterion erosive, the proportion for tropical rainfall is about 40 percent.

Furthermore, the average intensity of erosive rain is higher in the tropics: on the order of 60 mm hr^{-1} compared with 35 mm hr^{-1} for temperate erosive rain. The increase in intensity and in drop size is associated with an increase in the kinetic energy of the rain, of which erosivity is a function. The example given in Table 2.3 (Hudson, 1971) shows that for two comparable locations each receiving 1,000 mm of rain, erosivity of the rain at a tropical location could be about ten times greater than that at a temperate site. Rainfall intensity and energy load in a specific region, Northern Nigeria, is examined by Kowal and Kassam (1976).

■ 2.7.3 The actual amount of erosion that will occur during a storm of given erosivity will, as we know, be governed also by the *erodibility* of the soil (i.e., the potential of the soil to be eroded, as determined by its physical state under standard conditions), the angle and length of the slope, the management imposed on the crop and soil surface (e.g., crop population, mulching, etc.) and the conservation practices implemented (e.g., contour banks, strip cropping, etc.). The soil erosion aspects of farming systems will be examined more closely in later chapters, but Tables 2.4 and 2.5 exemplify the influence of slope, surface soil treatment, and soil fertility level (presumably acting through differences in the density of crop cover) on soil loss in two contrasting tropical locations in Africa.

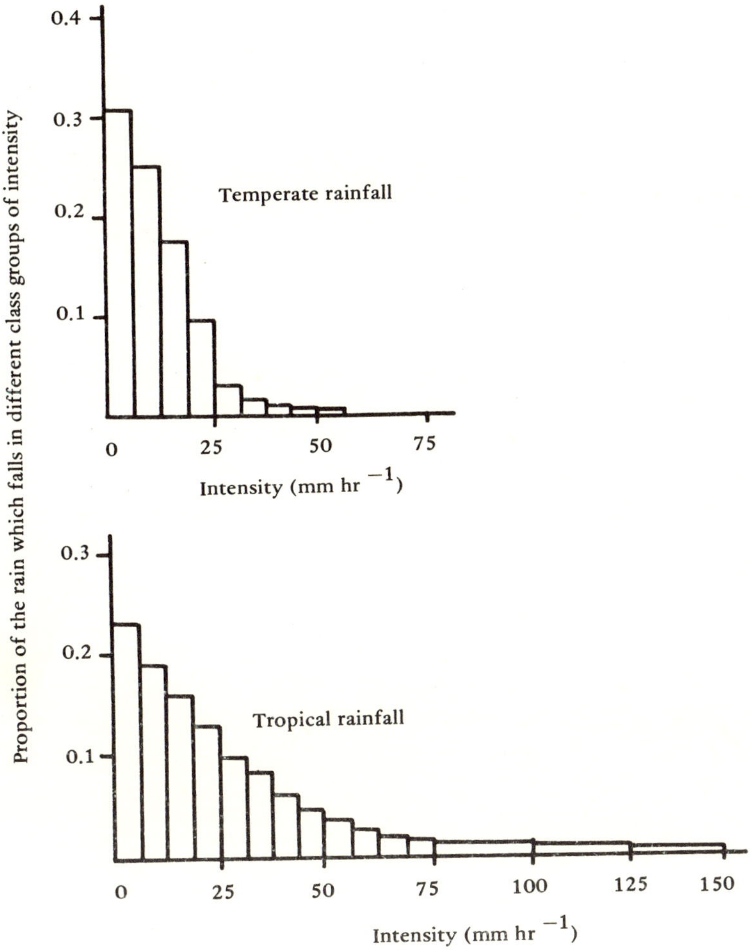

Figure 2.4. Frequency distribution of rainfall intensity in temperate and tropical regions (Hudson, 1971).

■ 2.7.4 Tables 2.4 and 2.5 are merely isolated specific examples from an infinite range of tropical cropping situations. To move to the other extreme, some highly generalized comments on water

TABLE 2.3
Erosivity of rain

	Temperate location	Tropical location
a. Amount of erosive rain in 1,000 mm rainfall (mm)	50	400
b. Kinetic energy of erosive rain (j m^{-2} mm^{-1})	24	28
c. Erosivity (a × b) (j m^{-2})	1,200	11,200

TABLE 2.4
Soil erosion from cornfields of differing slope and fertility level. Rhodesia

Fertility level	Slope (%)	Corn yield (t ha^{-1})	Soil loss (t ha^{-1})
High	3	8.43	2.24
	4.5	8.38	2.56
	6.5	8.24	2.68
Medium	3	4.08	3.56
	4.5	4.61	4.90
	6.5	5.00	7.60

SOURCE: Hudson (1971).

TABLE 2.5
Runoff and soil erosion after 64 mm of rain from fields of differing slope and rate of mulch application. Nigeria

Mulch rate (t ha^{-1})	Slope (%) 1	5	10	15	Mean
(a) *Runoff* (*mm*)					
0	46.0	59.8	37.7	59.8	50.8
2	1.3	16.6	15.2	14.0	11.8
4	0.4	1.5	3.6	3.3	2.2
6	0.0	0.7	1.9	1.8	1.1
(b) *Soil loss* (t ha^{-1})					
0	1.41	19.40	28.14	12.89	15.46
2	0.01	0.83	0.91	0.83	0.65
4	0.00	0.16	0.11	0.31	0.19
6	0.00	0.05	0.03	0.08	0.04

SOURCE: IITA (1974).

erosion and erosion control in relation to the broad categories of farming system (to which there are numerous exceptions) are given below:

a. In shifting cultivation, it is rare for soil conservation measures to be taken. Further, such cropping systems are often found on marginal sloping land liable to erosion. On the other hand, erosion potential is mitigated by the limited period of time for which any one cropfield is in the arable condition. During the fallow or recovery phase, liability to erosion is reduced by the developing vegetation cover.

b. Liability to erosion is greatest in intensive and semi-intensive rainfed annual cropping systems. Slopes are highly variable but the land surface is in an open arable condition either permanently or for a high proportion of time. Unhappily, in the tropics erosion control measures are relatively uncommon, except in areas where resource-conserving farming practices have evolved over many years in response to heavy pressure on land, or where extension of modern conservation technology has been effective.

c. Irrigated annual cropping systems are in most circumstances inherently soil-conserving. Although no specific control measures may be imposed, the surface geometry necessary for controlled water supply and drainage demands gentle slopes. This is certainly true of wet rice systems, with their near-flat fields and carefully constructed terraces. However, there are examples of poorly designed irrigation systems for upland crops in the tropics that are subject to erosion.

d. Land under perennial crops is least liable to erosion because of the protective effect of the permanent canopy. Furthermore, the capital resources of large-scale perennial crop enterprises are such that permanent conservation measures, i.e., contour terraces, are the rule rather than the exception. In cropfields of mixed annuals and perennials, e.g., the "garden" systems of Indonesia (see chapter 16), the tree canopy helps to protect the cropped area below.

■ 2.8 The storage component

■ 2.8.1 In the hydrological cycle of the tropical crop farm there is, of course, a natural storage component in the water held in the crop root zone. This store is continuously depleted by evapotranspiration and by percolation and is intermittently augmented by rainfall or irrigation. The magnitude of the available water store in the root zone will influence the performance of rainfed crops during dry spells, the required frequency of water application to irrigated crops, and under rainfed conditions can, as we have seen, have a marked effect on crop performance after the rains have ended. In some situations, drainage accumulated in the subsoil may be recovered for use within the boundaries of the farm by lifting or pumping from wells.

In addition, measures may be taken to extend storage in the crop root zone by water retention practices or structures. One of the simplest of these is a reduction of evaporation and runoff through the application of a surface mulch. Experience suggests that mulching is most effective in tropical regions of relatively high and frequent rainfall, e.g., in southern Nigeria. In India, in more arid summer-rainfall climates with longer and more intense within-season dry spells, such measures have proved less beneficial. An example of a simple water-retention structure is the broad contour bund commonly in use in India. The contour bund is primarily to control erosion, but it also acts to retain water on the upper side of the bund (though, as Kampen, 1974, has pointed out, its effectiveness as a water-conserving measure is questionable).

The traditional response in the development of water retention structures in the tropics has been the creation of a series of stepped terraces, with or without a surrounding narrow bund. The ultimate in control structures under conditions of high runoff and steep topography is seen in the meticulously terraced hillsides of Java and the Philippines, where in each near-flat rice field the required depth of water is stored and the excess drained through a small stop into the field immediately below.

However, the near-flat terraced field with its retaining bund is not as suitable for upland crops as it is for rice because of drainage problems within the field, and one central problem of the rainfed culture of upland tropical crops is to devise practical low-cost structures for the controlled disposal of runoff and, if possible, its retention on the farm for subsequent crop use. Alternative uses for water stored in this manner are for cropping in the no-rain period, i.e., for full dry-season irrigation, or for supplementary irrigation of rainfed crops during drought spells in the rain period. The work of Kampen and his associates at

ICRISAT (1974) in developing small cropped catchments with graded furrows, waterways, and small on-farm storage ponds, from which water may be released for gravity irrigation below or pumped back on to the catchment, points the way to future developments.

■ 2.8.2 Surface storage of runoff water and its subsequent use for cropping is, of course, a description of an irrigation scheme (or one type of irrigation scheme, since in others water is recovered from below-ground storages). However, in this book we are concerned with farming systems and, from our definition, with the patterns and processes of the individual production unit. Hence while it is appropriate to consider water retention and use on the farm, a detailed treatment of the operation of irrigation structures and schemes serving a group of farms or a large area is outside the scope of the book. Some of the problems associated with water supply to the individual farm from large-scale irrigation schemes are dealt with in chapter 14.

REFERENCES

Basinski, J. J. 1974. *Weather analysis and its application to production strategy in areas of climatic instability*. Rome: FAO. International expert consultation on the use of improved technology for food production in rainfed areas of tropical Asia.

Cochemé, J. 1968. Agroclimatology survey of a semi-arid area in West Africa south of the Sahara. In: *Agroclimatological methods*. Proceedings of the Reading Symposium. Paris: UNESCO.

Cochemé, J., and Franquin, P. 1967. An agroclimatology survey of a semi-arid area in Africa south of the Sahara. World Meteorological Organization. Technical note no. 86.

Fitzpatrick, E. A., and Nix, H. A. 1969. A model for stimulating soil water regime in alternating fallow-crop systems. *Agricultural Meteorology* 6:303.

Hudson, N. 1971. *Soil conservation*. Ithaca, N.Y.: Cornell University Press.

Indian Council of Agricultural Research. 1970. *Proceedings of the symposium on cropping patterns in India*. New Delhi, India: ICAR.

International Institute of Tropical Agriculture. 1974. *Annual report*. Ibadan, Nigeria.

Kampen, J. 1974. *Water conservation practices and their potential for stabilizing crop production in the semi-arid tropics*. Rome: FAO. International expert consultation on the use of improved technology for food production in rainfed areas of tropical Asia.

Kowal, J. M., and Kassam, A. H. 1976. Energy load and instantaneous intensity of rainstorms in Samaru, Northern Nigeria. *Tropical Agriculture* (*Trinidad*) 53:185.

Lockwood, J. G. 1974. *World climatology. An environmental approach.* London: Edward Arnold.

McCown, R. L. 1973. An evaluation of the influence of available soil water storage capacity on growing season length and yield of tropical pastures using simple water balance models. *Agricultural Meteorology* 11:53.

Ritchie, J. T. 1973. Influence of soil water status and meteorological conditions on evaporation from a corn canopy. *Agronomy Journal* 65:893.

Ritchie, J. T., and Burnett, E. 1971. Dryland evaporative flux in a subhumid climate: II. Plant influences. *Agronomy Journal* 63:56.

Trewartha, G. T. 1968. *An introduction to climate.* 4th ed. New York: McGraw Hill.

Westphal, E. 1975. *Agricultural systems in Ethiopia.* Wageningen, Netherlands: Centre for Agricultural Publishing and Documentation.

Chapter 3

General Energetic Background

■ 3.1 Solar energy to plant energy

■ 3.1.1 The theoretical maximum rate of carbohydrate production per unit area through photosynthesis has been calculated by Loomis and Williams (1963) as 710 kg ha^{-1} day^{-1}, assuming a solar energy input equivalent to the average radiation during a tropical summer growing season. If maintained year-round, this would give a yield of about 260 t ha^{-1} yr^{-1} of dry matter and would represent a solar energy use efficiency of 5.3 percent of total radiation or 12 percent of visible radiation.

This theoretical value assumes a photosynthetically active crop canopy intercepting all radiation, water and nutrients nonlimiting, and the crop free of disease and pests. Such a combination of circumstances rarely occurs in the real world except for brief periods. The maximum short-term efficiency of solar energy use recorded experimentally is between 4 and 5 percent of total radiation, giving a growth rate of 500 to 540 kg ha^{-1} day^{-1} DM over periods of 2 to 3 weeks. Such rates have been achieved with pearl millet in North Australia, corn in Japan and the U.S.A., and Sudan grass in the U.S.A (Loomis and Gerakis, 1975). For a complete year, the maximum recorded yield is 86 t ha^{-1} or 235 kg ha^{-1} day^{-1} from elephant grass in Central America.

All these figures are for aboveground growth only and would have been higher had root growth also been measured. However, they represent total dry matter yield, including leaves, stems, etc., not directly consumable by man. The highest grain yields recorded from a single crop are 21 and 16 t ha^{-1}, from corn in Michigan and rice in southern

Australia, respectively. Although in the tropics comparable yields for single grain crops have not been achieved, tropical temperature regimes permit under irrigation the growing of up to three cereal crops a year, and a total annual yield of 26 t ha^{-1} of paddy rice has been recorded at IRRI in the Philippines. This represents a utilization of about 0.5 percent of total radiation.

In practice, there are few areas in the world where three rice crops are grown each year under high-fertility conditions and strict control of pests and disease. National average yields of paddy rice in Southeast Asia, grown predominantly under one-crop-a-year systems, are of the order of 2 t ha^{-1}. For rainfed cereals in the semi-arid summer rainfall tropics the general level of yield is 0.5 to 1 t ha^{-1}, equivalent to only 0.01 to 0.02 percent solar energy conversion.

■ 3.1.2 Crop yields have hitherto been presented in terms of dry matter, but in order to evaluate the energetics of cropping systems they must be expressed in energy terms. The average energy value of a range of common tropical food crops is given in Table 3.1 (de Vries et al., 1967).

TABLE 3.1
Energy values of food crops

Cereals		Noncereals	
(MJ kg^{-1} air-dry)		(MJ kg^{-1} fresh weight)	
Rice	14.8	Cassava	6.3
Wheat	14.4	Sweet potato	4.8
Corn	15.2	Yam	4.4
Sorghum	14.9	Taro	4.7
		Plantain	5.4

Values for grain legumes are close to those for cereals, but are somewhat higher for oilseeds owing to the high energy value of the oil. A good average value for air-dry grains is about 15 MJ kg^{-1}.

■ 3.1.3 The above figures are for harvested economic yield, not all of which is edible. Whereas the whole of a grain of wheat or a kernel of corn may be consumed by man, paddy rice is milled and the outer layers of root and tuber crops removed before they are eaten. The edible portion of rice may be less than 60 percent; for root and tuber crops it is between 80 and 90 percent. However, this reduction does not necessarily represent an energetic loss to the farm system if the portions not eaten by man—rice hulls, tuber peelings, plantain skins, etc.—are utilized by farm stock.

Table 3.2 illustrates the general expected yield of edible energy obtainable under normal tropical farming conditions from a group of cereal and noncereal carbohydrate crops. The yields are three-year averages from FAO statistics. Edible energy output is expressed both per unit area and per unit area per unit time according to the average growing period for the crop. The final column represents edible energy per hectare per day from maximum-yield experiments, indicating the potential for improvement.

From the table it appears that, for a single crop, the average edible energy per hectare from noncereals is substantially higher than that from cereals. On the other hand, when the period of growth is taken into account, the distinction is no longer apparent. Clearly, however, the yield potential per unit time of short-season crops can only be fully realized if growing conditions—an extended season or irrigation—permit more than one crop a year.

■ 3.2 Seasonal variation in radiation and day length

■ 3.2.1 *Radiation.* In the wet tropics near the equator seasonal variation in total daily radiation is small, not only by virtue of the small changes in sun angle and length of day, but also because there are no distinct periods of clear and cloudy skies. In summer-rainfall climates, average daily radiation is not necessarily greater in summer than in winter, in spite of higher sun angles and longer days, because of the greater summer cloud cover. Table 3.3 gives radiation data for midsummer and midwinter months at eleven tropical locations; in three nonequatorial instances winter values are higher than summer values.

Although the general level of wet-season radiation in the tropics is less, for example, than that of a Mediterranean climate summer, it is normally adequate if not optimal for crop growth, though intermittent periods of prolonged cloudiness can delay maturation and inhibit reproductive development. The latter effect has been noted in cotton. Summer and winter radiation levels are important in rice-growing systems in summer-rainfall climates where irrigation permits dry-season cropping. Rice is commonly regarded as the characteristic cereal of the wet tropics; it is better thought of as the only cereal that can successfully be grown in low-lying areas in the wet tropics, or in the wet season of mon-

TABLE 3.2

Edible energy production per unit area and per unit time of some common tropical food crops

Crop	Average tropical yields ($t\ ha^{-1}$)*	Energy value ($MJ\ kg^{-1}$)*	Percent edible energy	Edible energy per unit area ($MJ\ ha^{-1} \times 10^3$)	Average growth period of crop (days)	Edible energy per unit area per unit time ($MJ\ ha^{-1}\ day^{-1}$)	Edible energy per unit area per unit time *Maximum*** ($MJ\ ha^{-1}\ day^{-1}$)
Rice	2.0	14.8	70	20.7	150	138	740
Wheat	1.2	14.4	100	17.3	120	144	460
Corn	2.1	15.2	100	31.9	135	236	840
Sorghum	1.0	14.9	90	13.4	135	99	480
Cassava	9.1	6.3	83	47.6	330	144	1,050
Sweet potato	6.5	4.8	88	27.4	135	203	760
Yam	8.0	4.4	85	29.9	280	107	—
Taro	5.8	4.7	85	23.2	120	193	—
Plantain	21.1	5.4	59	67.2	365	184	340

SOURCE: de Vries et al. (1967).

*Cereals: air-dry weight; non-cereals: fresh weight.

**From maximum yield experiments.

soon climates, because of its tolerance to waterlogging. In monsoon climates, if other conditions are not limiting, higher yields are obtained in the high-radiation winter dry season. Low temperature may extend crop maturation by 25 to 30 percent, but at IRRI in the northern Philippines, for example, maximum dry-season rice yields are on average about 40 percent higher than maximum wet-season yields.

■ 3.2.2 *Day length.* The seasonal variation of day length in the tropics, though narrower in amplitude than that of temperate regions, is of agricultural importance, since a substantial proportion of unimproved

TABLE 3.3
Average daily solar radiation in midsummer and midwinter months at tropical locations (jcm^{-2} day^{-1})

N. latitudes: S. latitudes:	Dec. June	June Dec.
Puerto Rico (N)	1,590	2,140
Palmira, Colombia (equat.)	1,510	1,420
Singapore (equat.)	1,590	1,760
*Colombo, Sri Lanka (N)	1,760	1,260
*Lagos, Nigeria (N)	1,680	1,260
Hawaii (N)	1,340	2,010
Havana, Cuba (N)	1,300	1,680
*El Salvador (N)	1,630	1,260
Trinidad (equat.)	1,590	1,720
Townsville, Australia (S)	1,680	2,300
Katherine, Australia (S)	1,800	2,010

SOURCE: Cooper (1970).
*Midwinter values higher than midsummer.

tropical crop cultivars is photosensitive. The day length range may be small, but local cultivars have evolved over the years with small but agronomically significant differentiation in day length response. The crops that are photosensitive are broadly speaking "short-day" plants, but they show a range of sensitivity; flowering time is also modified by temperature, and crop behavior cannot always be explained on the basis of a simple "critical day length" response.

The process of local cultivar evolution toward a specific photoperiod response is seen most clearly in savannah climates, since this climate pattern is typical of the higher latitudes with an appreciable summer-winter day length range. The maturation pattern of photosensitive rainfed crops that have evolved at a specific location shows adaptation to the average end-date of the growing season, as determined by P and E_t, so that in a normal season they complete their maturation processes largely on stored water (see Bunting and Curtis, 1968, for the example of sorghum in West Africa). In this way cultivars escape the disease hazards of maturing while the rains are still in progress, but at the expense of yield, since available soil water for grain-filling is not always adequate.

The advantage of a photoperiod response is that maturation and harvest are largely independent of time of sowing. If sowing is delayed, as it often is because of a late start to the rains or because of the peak of labor demand at sowing-time, some yield will be obtained even if crop development is telescoped. On the other hand, a late-sown nonsensitive cultivar of fixed maturation period may not complete reproductive development before the rains end. Furthermore, with photosensitive cultivars the labor peaks at harvest are more predictable.

Modern trends in crop breeding are, however, toward photoinsensitivity; to give greater flexibility to planting time and hence to cropping patterns in general, to permit breeding for early maturity and hence a potential increase in cropping index, and to allow advantage to be taken of extension to the rainfed growing season through irrigation development projects. On the other hand, photoinsensitive cultivars sown early in the wet season may mature during periods of high rainfall probability, thus increasing the risk of grain disease. At ICRISAT in Hyderabad, for example, resistance to grain mold diseases is an essential aim in the breeding of quick-maturing photoinsensitive sorghum and pearl millet.

■ 3.3 Efficient use of solar radiation

■ 3.3.1 We are accustomed to monoculture—one field, one crop—as the norm in modern mechanized agriculture, and we are familiar with the efficiencies and inefficiencies of solar radiation interception by crops at different phases of their development and by crops of different canopy geometry. It is the tyranny of mechanization that enforces monoculture; the nonmechanized tropical crop farmer is not so bound, and he frequently takes advantage of this by sowing mixtures of crops. By sowing more than one crop, perhaps at different times, of different plant habit and of different canopy development pattern, he maximizes the use of solar radiation (and soil water) for crop production. As we shall see later, however, this by no means is the only advantage, nor the most important one, to be gained by growing more than one crop at the same time on the same area. The subject of multiple cropping has been very thoroughly reviewed recently (American Society of Agronomy, 1976).

■ 3.3.2 *Shade crops.* Some perennial tree crops, e.g., cocoa, coffee, tea, are shade-tolerant and in fact, if other environmental conditions are not optimal, may be physiologically more robust when grown in partial shade. We are not here concerned with perennial crop systems nor with perennial shade trees, but the shading of one crop by another is important in some mixed crop systems. Thus, the intercropping of tall arable crops such as bananas or plantains in young cocoa or coffee not only utilizes land more fully while the trees are small, but also provides lateral shade. In many wet tropical shifting cultivation systems, shade-tolerant and shade-intolerant crop species are grown together, ensuring maximum solar energy use.

■ 3.4 Human food chains

■ 3.4.1 In the production of crops for direct consumption by man, though the primary conversion from solar energy to plant energy is inefficient, the energy losses between the primary crop product and the edible portion are relatively small, as shown in Table 3.2. However, direct consumption of crop products is only one of the human food chains, which can be grouped into four main classes (Duckham and Masefield, 1971):

a. direct consumption of arable crops by man, as above;
b. arable crops fed to livestock, the products of which are consumed by man: e.g., poultry fed on corn, pigs on sweet potatoes, beef cattle on grain sorghum, etc.;
c. forage consumed by ruminant livestock, the meat of which is consumed by man: e.g., grassfed beef, lamb, etc.;
d. forage consumed by ruminant livestock, the milk from which is consumed by man: e.g., cattle or buffalo milk.

■ 3.4.2 The inefficiencies in energy conversion of food chains (b), (c), and (d) are much greater than those of (a). Data are not readily available for tropical crops and livestock, but the figures from Duckham and Masefield (1971) given in the following list exemplify the comparative energy losses from chain (a), represented by Irish potatoes, and chain (c), represented by beef raised intensively on pastures. Both sets of data are from temperate situations. The figures quoted are the partitioning of total energy of the primary product, in (a) the whole potato plant and in (b) total forage.

a.	*Direct consumption of potatoes*	
	Nonconsumable tops and roots	24%
	Post-harvest storage loss	14
	Household waste	14
	Net human food	48
b.	*Meat from grassfed cattle*	
	Uneaten forage (stubble and roots)	16
	Uneaten forage (leaf material)	22
	Faeces and urine	22
	Animal metabolic and conversion losses	31
	Household waste	3
	Net human food	6

Thus in energy terms the efficiency of the step from the primary products of photosynthesis to useful food for man is eight times greater for potatoes than for beef. (The energy costs of packaging, storage, transport, and retail distribution are excluded.) Duckham and Masefield also give some comparative figures on the efficiency of the total conversion process from solar energy to useful food:

a.	Intensive rice production	0.17%
b.	Pig meat, pigs on arable crop feed	0.03
c.	Lamb meat, lambs on pasture	0.01
d.	Cow milk, cows on pasture	0.05

■ 3.5 Human energy requirement and expenditure

■ 3.5.1 In modern mechanized crop production systems, the human energy component, evaluated in muscular energy expended or food energy consumed, of the total nonsolar energy input is dwarfed by the high direct and indirect fossil fuel requirement. Thus for corn production in the U.S.A., human energy input in 1945 contributed 1.3 percent of the total, but by 1970 the proportion had fallen to 0.17 percent (Pimentel et al., 1973). In tropical farming systems, mechanization and the substantial use of fertilizers is infrequent and the nonsolar energy input is largely human energy alone or human plus draft animal energy. It is therefore essential to grasp the general energy principles of human body maintenance and work.

The energy requirements of humans vary according to body weight, the temperature of the environment, and the level of exertion. There is considerable variation in average body weight between races and naturally between ages and sexes, with consequent variation in energy requirement. Approximate average adult male weight for different regions of the world are given by Clark and Haswell (1970) as U.S.A., 70 kg; Europe, 65 kg; Africa, 55 kg; Latin America, 55 kg; and Southeast Asia, 50 kg.

The influences of age and sex on energy requirements are seen in Table 3.4 (FAO, 1958) for people of 50 kg average adult male weight in a 25°C environment—Southeast Asians, for example. For such a regional group, the average energy requirement, weighted by the age and sex distribution of the population, is 8.4 MJ day^{-1}.

TABLE 3.4
Energy requirements in a 25°C environment for people of average adult male weight = 50 kg

Age (yrs)	Male (MJ day^{-1})		Female (MJ day^{-1})
0– 1		4.7	
2– 3		5.0	
4– 6		6.6	
7– 9		8.2	
10–12	9.7		9.3
13–15	12.0		10.1
16–19	11.6		7.4
20–29	10.2		7.1
30–39	10.0		6.9
40–49	9.6		6.6
50–59	8.9		6.1
60–69	8.1		5.6
70 +	7.1		4.9

■ 3.5.2 In estimating the daily energy requirements of farm families, we need to know the level of energy expenditure for different activities. Figures for African adult males weighing 55 kg, in a tropical environment, are given by Philipps (1954):

Field work	1.05 MJ hr^{-1}
Walking without load	0.50 MJ hr^{-1}
Sedentary	0.33 MJ hr^{-1}
Sleeping	0.26 MJ hr^{-1}

Naturally, energy expenditure varies with the work task being performed:

Hoeing	1.15 MJ hr^{-1}
Weeding	1.13 MJ hr^{-1}
Clearing bush	1.56 MJ hr^{-1}
Felling trees	2.11 MJ hr^{-1}

■ 3.5.3 The figures immediately above are for continuous activity and all are greater than the average value for field work of 1.05 MJ hr^{-1}. Clearly the harder the work, the more frequent and prolonged the rests; hence there is a tendency for energy expenditure over a working day to average out at the same figure regardless of the task. Indeed, Haswell (1973) presents data which suggest that over a full working day (length not specified) average energy expenditure, allowing for short breaks and long rests, is remarkably constant whatever the task being performed and averages only about 0.7 MJ hr^{-1}.

Although we tend to think of farmers working hard all day, surveys in tropical developing countries (Haswell, 1973) indicate that the working day averaged over a full year is often only about 4 to 5 hours. This is often largely due to the highly seasonal nature of farming activity and consequent enforced idleness during the no-crop period. It is, of course, likely to vary between farming systems, climates, and perhaps between ethnic groups. If we take a typical example of an African adult male working hard on the average for 5 hours a day and walking for a further 2 hours, his daily rate of energy expenditure might be:

5 hours working at 1.05 MJ hr^{-1}	5.2 MJ
2 hours walking at 0.50 MJ hr^{-1}	1.0 MJ
8 hours sedentary at 0.33 MJ hr^{-1}	2.6 MJ
9 hours sleeping at 0.26 MJ hr^{-1}	2.3 MJ
Total	11.1 MJ

■ 3.5.4 In general, it appears that the average energy requirement of adult males in the tropics is within the range 10 to 12 MJ day^{-1}. For the population as a whole, this would imply an average range of

8 to 10 MJ day^{-1}, remembering not only that the other classes of the population weigh less, but also that a proportion of the population would not be working.

These estimates tally reasonably well with the results of a major survey in the 1950s of average energy intake, undertaken by the USDA and quoted by Clark and Haswell (1970), which yielded the following ranges:

Tropical Africa: 9.1 (Tanzania) to 11.2 (Nigeria) MJ day^{-1}
Tropical Asia: 8.6 (India) to 9.2 (Thailand) MJ day^{-1}
Tropical America: 7.8 (Haiti) to 10.2 (Costa Rica) MJ day^{-1}

There are, as we would expect, substantial differences in energy intake between regions within countries in accordance with local levels of food production and consumption. Clark and Haswell give some comparative data from environments in India:

Area of good loam soils, irrigated,	9.0 MJ day^{-1}
Area of black clay soil, rainfed,	8.4 MJ day^{-1}
Area of poor sandy soil, rainfed,	7.8 MJ day^{-1}

Furthermore, for marginal subsistence farmers in climates where the main food supply is harvested at one single time of year, there is often a seasonal variation in food intake, showing a maximum immediately after harvest and declining as the store of food diminishes. Table 3.5, from Clark and Haswell, records data collected in the Sudan illustrating this seasonal variation.

TABLE 3.5
Seasonal variation in food intake. Sudan

	Energy intake (MJ day^{-1})	Protein intake (g day^{-1})
After main harvest	11.6	60
In dry season	10.0	56
In early wet season	8.1	33

It will be realized that in such summer-rainfall climates the farm labor force is working hard growing the next crop at a time when food supply is at its lowest. It is also noteworthy that the decline in protein intake from harvest onwards is even greater than that of energy intake. The topic of seasonal hunger is discussed by Johnston (1958).

■ 3.5.5 In the foregoing section the word protein was mentioned for the first time, and it is reasonable to query whether the evaluation of food production systems in energy terms above does not ignore a vital feature of the "food equation," in that protein intake may also be a major limiting factor to human productivity and health in many tropical regions.

The short answer is, of course, that although we can measure food output in terms of both energy and protein, we can only measure human input in terms of energy. If protein intake is in excess of normal requirement then it will be utilized in work or other activity according to its energy value; if protein intake is deficient then it may have the result of limiting activity and hence energy input into farm work. But we cannot express human work capacity in terms of protein requirement.

In addition, the immediately postwar view that the primary deficiency of malnourished people in tropical regions is protein is now seriously questioned, and more attention is now being paid to energy. Malnutrition is now generally referred to as protein-calorie deficiency (Waterlow and Payne, 1975). With increasing medical knowledge, standards of human protein requirement have been steadily lowered, and are now only about two-thirds of those proposed in 1948. Our lack of certainty on such an important question arises from the fact that protein and energy deficiencies tend to occur simultaneously and because it is ethically impossible to conduct sustained experiments in human malnutrition to distinguish the two elements.

This is an enormous subject to dismiss in a few sentences, but the brief mention is perhaps needed in order to justify confining our attention here to the energetic aspects of food output and consumption. In certain tropical regions, in certain years and certain seasons within years, and for certain age and sex classes, protein deficiency may be of primary importance. However, it is broadly true to say that for tropical agricultural peoples on a diet of which the main energy component is cereal grain, plus a certain amount of grain legume and vegetable supplemented with a little meat or fish, protein intake is adequate; for such people it is the total energy balance that is of major importance.

■ 3.6 Energy output/input ratio

■ 3.6.1 In recent years, particularly because of the marked increases in fossil fuel costs, much attention has been paid to assessment of the energetic efficiency or "caloric gain" of farming operations and farming systems: that is, the ratio of food energy output to energy input. Most of the research is concerned with mechanized agriculture in technologically advanced nations and has been extended from production systems to food systems as a whole; i.e., the energetics of the total process of bringing food to the dinner table. A very limited amount of work has been done on nonmechanized tropical crop production systems.

The earliest attempt was that of Black (1971). Using data from a number of sources, he estimated energy outputs and inputs for thirteen hand-cultivation cropping enterprises in the tropics and arrived at an average value for energy output/input of 17:1. The cropping systems were shifting or semi-intensive rainfed cropping. Black's data will be examined more closely in later chapters; at this point it is only necessary to clarify the basis of output/input estimates when they are applied to farming systems where the major proportion of input is human energy.

■ 3.6.2 Energy expenditure has two components: basal metabolic energy expenditure, which continues all the time whether the person concerned is working or not, plus the additional energy expended in work. The latter is *net energy expenditure*, the total is *gross energy expenditure*. The values given in section 3.5.2. for various farming tasks are gross energy expenditure. However, it is clear that if we wish to compare the energy output/input ratios of different cropping enterprises or farming systems, the appropriate denominator is net energy expenditure. This denominator was used by Black (1971) in his calculations, though he adopted the rather low figure of 0.63 MJ hr^{-1}. In subsequent examples in this book, unless otherwise stated, the standard rates of human energy expenditure are assumed to be:

Basal energy expenditure	0.25 MJ hr^{-1}
Net work energy expenditure	0.75 MJ hr^{-1}
Gross work energy expenditure	1.00 MJ hr^{-1}

On the other hand, if we wish to compute the total energy balance of the farm worker, the appropriate denominator for the calculation of output/input could be said to be his total gross energy expenditure, working, walking, sedentary, sleeping, etc., throughout the day.

Let us assume that our farmer or farm worker spends an average in the year of 5 hr day^{-1} in crop production tasks, during which time he expends 1.0 MJ hr^{-1} (0.75 work, 0.25 basal), and for the rest of his time is fairly idle, expending only 0.3 MJ hr^{-1} gross. This gives an average total daily energy expenditure of $(5 \times 1.0) + (19 \times 0.3)$, or 10.7 MJ day^{-1}.

If we estimate input solely on the basis of net energy expenditure while working it is $365 \times 5 \times 0.75$, or 1,369 MJ yr^{-1}; if we use total gross energy expenditure, working and nonworking, it is $365 \times 10.7 =$ 3,906 MJ yr^{-1}. If total food energy output were, say, 30,000 MJ, then the respective output/input ratios would be approximately 22:1 and 8:1.

■ 3.6.3 There are two further complexities. The first stems from the difficulties in defining what are and what are not crop production tasks. Clearly, clearing land, ploughing and hoeing are, and, less certainly, walking to and from cropfields. But what about processing crops for home consumption, or carrying them to the market to sell? The second complication concerns who are workers and who are not, for in a normal family farm there will be a full range across the ages and sexes from those who carry most of the burden to those who, because they are too young or too old, or are incapacitated, make no contribution to crop production.

It would appear from the above that there are two useful measures of human energy expenditure that can be used on the input side, and each has its function. If we wish to compare the energetic efficiency of different cropping enterprises or operations, say corn-growing against cassava-growing, then the appropriate estimate of input energy is the net energy expended by the workers in crop production. If, however, we are attempting to estimate the energy balance of a complete farm, perhaps representing a particular farming system that we wish to compare with another, then the relevant estimate of input energy is total gross energy expenditure by the farm family, whether working or not, plus any additional labor. For the first type of estimate it is necessary to define carefully what are and are not crop production tasks; for the second type it does not matter.

■ 3.7 The energetics of draft animals

■ 3.7.1 Draft animal energy is an important component of the energy balance of farming systems in tropical Asia, less so in tropical Africa and America. However, surprisingly little research or survey work has been done on the energetics of draft animals.

Odend'hal (1972) measured the mean daily energy intake of draft cattle in a monsoon climate area of India (rainfall 1,250 to 1,750 mm) as:

Adult males	91.1 MJ day^{-1}
Adult females	57.0 MJ day^{-1}
Immature cattle	37.9 MJ day^{-1}

Basal, net working and gross working energy expenditure by draft animals will be dealt with in chapter 10.

Another potentially important energy output from draft animals is their dung. In areas of the tropical world where both human and ruminant populations are dense and wood scarce, as in India, dung is used extensively as a fuel. The Indian Ministry of Food and Agriculture estimates that about two-thirds of the cattle and buffalo dung produced in India is so used (National Council of Applied Economic Research, 1965).

■ 3.7.2 In evaluating the energetics of draft animals within a farming system, there are two important points to bear in mind. First, although many tropical farmers with draft animals do not breed them, many others maintain a small breeding herd to provide replacements and perhaps some milk for the family or for sale. In any energy budget these nonworking animals have to be accounted for, as with the human beings. Or to use an analogy drawn by Revelle (1976), we can regard the draft animal as a "working machine" and the energy cost of maintaining a herd to replace it as being equivalent to the indirect energy cost of manufacturing a tractor.

The second point is less readily disposed of. In some areas of the Asian tropics, such as India, where ruminant and human populations are high, the proportion of uncropped land is small and the grazing pressure on it very heavy. Under such conditions, draft animals may subsist largely on crop residues and by-products. For example, in Odend'hal's study it was estimated that 94 percent of the energy intake came from rice straw, rice hulls, and sugarcane tops.

However, in areas of lower population, both human and ruminant, and hence characterized by a much higher proportion of uncropped land subjected to a lower grazing pressure, draft stock obtain the major proportion and in some instances all of their energy requirements from ill-defined sources: village commons, roadsides, irrigation canal banks, open forest, etc. The forage may or may not be within the physical boundaries of the farm; in either case it is scarcely considered as part of the normal farm "output." Furthermore, a variable amount of human energy has to be expended in harvesting and storing crop residues or in gathering natural forage. Clearly, estimating the energy cost of maintaining draft stock in such situations is difficult.

REFERENCES

American Society of Agronomy. 1976. *Multiple cropping*. Madison, Wisconsin: ASA Special Publication no. 27.

Black, J. N. 1971. Energy relations in crop production, a preliminary survey. *Annals of Applied Biology* 67:272.

Bunting, A. H., and Curtis, D. L. 1968. Local adaptation of sorghum varieties in northern Nigeria. In: *Agroclimatological methods*. Proceedings of the Reading Symposium. Paris: UNESCO.

Clark, C., and Haswell, M. 1970. *The economics of subsistence agriculture*. 4th ed. London: Macmillan.

Cooper, J. P. 1970. Potential production and energy conversion in temperate and tropical zones. *Herbage Abstracts* 40:1.

Duckham, A. N., and Masefield, G. B. 1971. *Farming systems of the world*. London: Chatto and Windus.

Food and Agriculture Organization. 1958. Calorie requirements. Nutritional Studies no. 15. Rome.

Haswell, M. R. 1973. *Tropical farming economics*. London: Longman.

Johnston, B. F. 1958. *The staple food economies of western tropical Africa*. Stanford, California: Stanford University Press.

Loomis, R. S., and Gerakis, P. A. 1975. Productivity of agricultural ecosystems. In: Cooper J. P., ed. *Photosynthesis and productivity in different environments*. London: Cambridge University Press.

Loomis, R. S., and Williams, W. A. 1963. Maximum crop productivity: an estimate. *Crop Science* 3:67.

National Council of Applied Economic Research. 1965. Domestic fuels in rural India. New Delhi: NCAER.

Odend'hal, H. 1972. Energetics of Indian cattle in their environment. *Human Ecology* 1:3.

Philipps, P. G. 1954. The metabolic cost of common West African agricultural activities. *Journal of Tropical Medicine and Hygiene* 57:12.

Pimentel, D., Hurd, L. E., Bellotti, A. C., Forster, M. J., Oka, I. N., Sholes, O. D., and Whitman, R. J. 1973. Food production and the energy crisis. *Science* 182:443.

Revelle, R. 1976. Energy use in rural India. *Science* 192:969.

de Vries, C. A., Ferwerda, J. D., and Flack, M. 1967. Choice of food crops in relation to actual and potential production in the tropics. *Netherlands Journal of Agricultural Science* 15:241.

Waterlow, J. C., and Payne, P. R. 1975. The protein gap. *Nature* 258:113.

Chapter 4

General Biogeochemical Background

■ 4.1 The major tropical vegetation formations

■ 4.1.1 In modern temperate crop farming systems, the natural vegetation often plays little or no part in the biogeochemical cycle. If the cropping pattern is one of a crop phase alternating with a grassland phase, the grassland will almost certainly be composed of improved sown species. However, in shifting cultivation and semi-intensive annual cropping systems in the tropics, the natural vegetation, or some successional phase of it, plays such an important role in biogeochemical processes that a brief account of its character is essential as background. The following very condensed survey is based largely on Eyre (1963).

■ 4.1.2 *Tropical rainforest* is associated with low altitudes and wet tropical climates normally with more than 2,000 mm rainfall. In America, this formation is centered on the Amazon basin and in Africa on the Congo basin. However, human occupation of Africa has a long history and many rainforest areas, particularly marginal ones, have been highly modified. In Asia, tropical rainforests may be found from East India to Northeast Australia, but many large areas have been cleared or greatly changed.

Structurally, rainforests are remarkably similar around the world. Nearly all the tree species are evergreen, with peak heights of 30 m; the shrub and ground layers are poorly developed. The forests are farmed largely under shifting cultivation. Since reassertion of the climax vegetation after clearing, burning, and cropping may take 50 to 100 years, the communities are, except in areas of low population, permanently in a successional phase. In areas where rainforests were formerly developed on fertile soils that have been subsequently densely settled, as in Sri Lanka or Java, the community has been replaced by cropfields and is now vestigial. In other areas, West Africa and Southeast Asia particularly, it has been replaced by perennial tree crops. Ruminant livestock are sparse since herbaceous species are scarce and insect and disease challenge severe.

■ 4.1.3 *Semi-evergreen seasonal forest* is associated with lowland or moderate-altitude summer rainfall climates with a distinct but not prolonged dry season. In America, the rainfall limits are approximately 1,000 to 1,500 mm, but in Asia the formation extends into higher rainfall regions. Semi-evergreen seasonal forest is geographically peripheral to the main forest zones. Much of the Indian subcontinent must have been semi-evergreen forest at one time, but through long and intensive cultivation the community has largely disappeared. There are few communities in Africa that can be classified as semi-evergreen forest, probably because under long human occupation most of the former areas have become savannah.

Average maximum height is about 20 m, and about 20 to 30 percent of the tree species are deciduous. Light penetration is greater than in tropical rainforest, and shrub and ground layers are generally better developed. Semi-evergreen forest land of low to moderate relief has been extensively developed for cropping at varying intensities, and areas of sharp relief are often used for shifting cultivation. The replacement of such communities by savannah grassland through human disturbance is frequently associated with an increase in livestock, mainly cattle.

■ 4.1.4 *Deciduous seasonal forest* is normally found at low to moderate altitudes in summer-rainfall climates with a dry season of at least five months and in rainfall zones of 600 to 1,000 mm. In America, the formation is well represented in the drier areas of Central America and the Caribbean. It has been greatly modified on the Indian subcontinent but is still found in Southeast Asia. Africa has vast belts of deciduous forest both in the southern tropics and in the sub-Sahara zone.

The average maximum height is 7 m; tree density is less than in other tropical forests. The relative proportions of tree, shrub and ground layer vegetation are strongly influenced by the degree of human interference through cultivation, fire, and grazing, and the boundary between deciduous forest and wooded savannah is often somewhat arbitrary. The deciduous forest zones are the home of semi-arid annual cropping systems, from shifting cultivation to intensive annual cropping, and of pastoral systems, which in Africa may be seminomadic.

■ 4.1.5 *Forest and savannah.* The term savannah is applied to a wide range of communities from treeless grassland to woodlands with an almost complete canopy cover, their common feature being a ground cover dominated by grasses or sedges. Although there is a general correspondence between the occurrence of savannah vegetation and the incidence of a climate with clearly defined wet and dry seasons, the dominant factor determining the continued existence of savannah, with some important exceptions, is regular burning. Figure 4.1 from Moss (1969) illustrates the ecological and anthropogenic factors governing the boundaries between forest and savannah.

■ 4.1.6 *Tall-grass savannah* is a general term for a community found extensively only in Africa, where it is known as "high grass-low tree" or "elephant grass" savannah. There are two main belts flanking the tropical rainforest and interdigitating with it. The incidence of this community and the sparse representation of semi-evergreen forest in Africa appear to be linked and both related to the effects of long human occupation. Often the local boundaries between tall-grass savannah and forest are very sharp: a self-perpetuating demarcation maintained by fire, which is confined to the cleared grassy area. Tall-grass savannahs are customarily utilized for semi-intensive rainfed annual cropping, which may or may not include ruminant livestock.

■ 4.1.7 *Mid-grass savannah* is a vegetation type referred to in Africa as "*Acacia*–tall-grass" savannah; other names are applied elsewhere. Since the average height of the grasses is 1 to 2 m, mid-grass is perhaps a more appropriate term. Such communities are found in zones of lower rainfall than the tall-grass savannahs. They occupy great belts across the northern African tropics and to the east and south of the forests. In America, the campos of Brazil and the llanos of the Orinoco basin fall into this category. In Asia, areas are found in the Indian Deccan and Burma, and in East Indonesia and North Australia the formation is well represented. Apart from providing the base for annual cropping systems of varying intensity, these savannahs also support extensive pastoral systems, sedentary in America but often seminomadic in Africa.

■ 4.1.8 *Edaphic savannahs.* Other savannah communities found in lower rainfall zones, where cropping is not feasible, need not

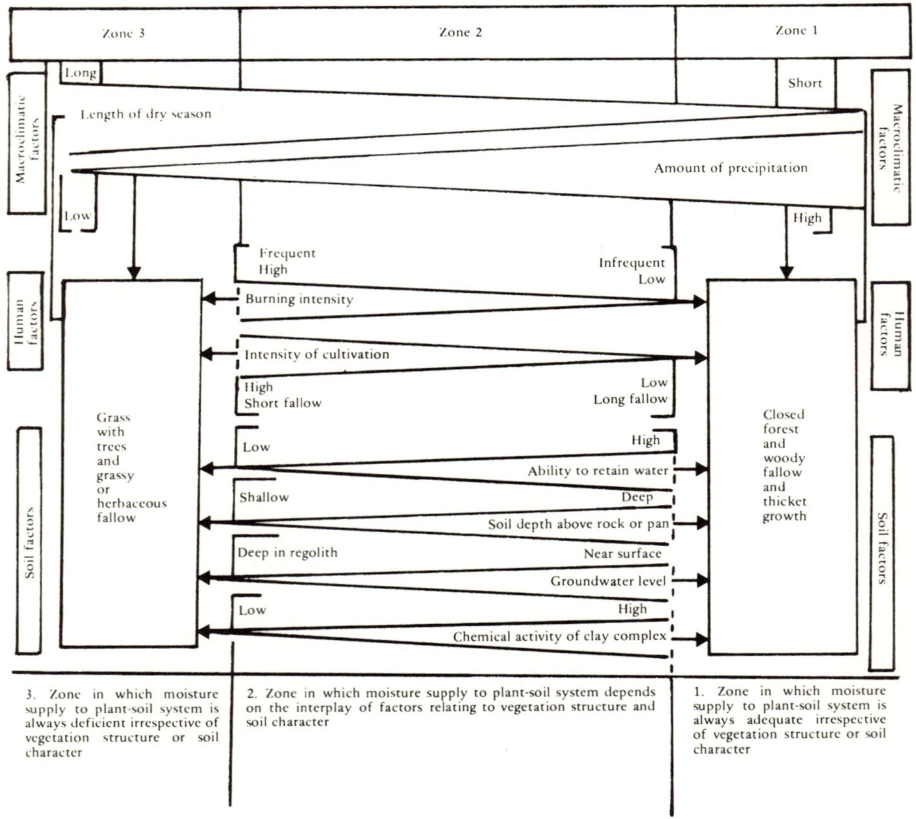

Figure 4.1. Factors involved in determining the boundary between savannah and forest (Moss, 1969).

concern us, but mention should be made of edaphic savannahs, the character of which is determined largely by soil and topographical factors. Edaphic savannahs, which are normally treeless grass or sedge communities, are associated with heavy soils, poor drainage, and liability to waterlogging or flooding. The llanos is by some regarded as edaphic savannah; these communities are also extensive in North Australia. Where such grasslands occurred originally on relatively fertile heavy soils, as in India and the Sudan, they have been extensively exploited for cropping, but under conditions of low fertility and high flood incidence they are largely utilized for grazing.

■ 4.2 The biogeochemical cycle: inputs

■ 4.2.1 Having briefly described an important section of the biogeochemical resource, the natural vegetation, we may now turn to the main concern of this chapter, which is to review in summary the components of the biogeochemical cycle in tropical farming systems, beginning with the inputs. These may be classified as:

a. Release of available nutrients from the soil or vegetation store.
b. Fixation of atmospheric nitrogen via crop legumes.
c. Other processes of atmospheric nitrogen fixation.
d. Nutrients in rainfall or run-on water.
e. Nutrients in stockfeed brought on to the farm.
f. Fertilizer nutrients.

It is debatable whether the release of nutrients from the soil store —item (a) in the above list—should be regarded as an input or as a component of the internal transformations or cycles within a farming system. It depends on how the boundaries of the "system" are defined. In the sense that the system represents the actual operations and cycles associated with cropping, then the release of nutrients into the system from an otherwise unavailable store is an "input."

■ 4.2.2 *Release of nutrients.* In view of the very high proportion of tropical cropland that does not receive fertilizer, neither purchased mineral fertilizer nor animal manure, it is the rate of release of available nutrients from the soil store and their fixation or immobilization that exercises a dominant influence on productivity. Furthermore, the rationale of those farming systems involving a fallow break, in which the recovering vegetation is cleared and burned before a cropping phase, is the release of nutrients from the vegetation store through fire.

Changes in the availability of soil nutrients other than nitrogen and sulfur are largely a characteristic of the local soil and climate and are not readily modified by farming practice except by vegetation accumulation and burning. An important exception to this is the effect of flooding in rice fields, which increases the availability of soil phosphorus (Patrick and Mahapatra, 1968).

On the other hand, the basic pool of nitrogen and sulfur is in the soil organic fraction, and their availability is broadly influenced by the cropping pattern. The key labile element is nitrogen, the supply of which as nitrate or ammonium ions is a function of the rate of mineralization of soil organic matter. This may be expressed in terms of the *mineralization coefficient*, or the amount of nitrate-N formed in one cropping season as a percentage of the total nitrogen in the root zone (Wetselaar, 1967a). It is commonly about 2 to 5 percent in a well-drained soil.

In the release of nutrients from the vegetation store by cutting down, allowing to dry, and burning, nitrogen and sulfur are lost to the atmosphere as oxides, but phosphorus, potassium, calcium, and magnesium remain in the ash, a ready source of available nutrients for the crop.

The most important of the several processes of immobilization and fixation is the capacity of tropical soils to fix phosphate. This varies widely of course, but many of the Oxisols, Ultisols, and Andept-Inceptisols are capable of immobilizing large amounts of added fertilizer phosphorus (Kamprath, 1973).

■ 4.2.3 *Legume fixation of nitrogen.* Although the net annual increment of nitrogen in the soil under tropical pasture legumes may be high, this is not true for short-season annual crop legumes. In general, it is more realistic to regard crop legumes as being nondepleting rather than restorative of soil nitrogen. The main reasons for this appear to be:

a. The amount of nitrogen fixed is broadly related to the duration of the crop, and annual legumes occupy the ground for relatively short periods of 2 to 5 months.
b. Much of the crop nitrogen is removed from the field at harvest. The major proportion of final plant N will be in the grain; furthermore, residues are frequently not consumed *in situ* by

stock and thereby returned to the soil in dung and urine, but are removed and fed elsewhere or burned.

c. The operations of plowing and cultivating increase the rate of mineralization, and a proportion of the mineral nitrogen so formed may, on well-drained soil, be leached out of the root zone.

d. Soil nitrogen made available at the start of the growing season may reduce the amount of atmospheric nitrogen fixed since the legume may take up mineral nitrogen preferentially.

A striking example of the long-term effect of crop legumes comes from Katherine, North Australia, where on a free-draining soil in a 900 mm summer rainfall climate seventeen years of continuous peanuts actually reduced total nitrogen in the root zone, though leached nitrate nitrogen accumulated in the subsoil.

■ 4.2.4 *Other nitrogen fixation.* The most important proven source of fixed atmospheric nitrogen in tropical farming situations, other than that from the legume-*Rhizobium* symbiosis, is that associated with flooded rice fields (Yoshida and Ancajas, 1973; Watanabe et al., 1977). Fixation is of two types: by photoautotrophic algae in the floodwater and by heterotrophic bacteria in the soil around rice roots. Increments of up to 40 kg ha^{-1} N per rice crop have been estimated (Firth et al., 1973). At IRRI, twenty-three rice crops have been grown over eleven years without any apparent decline in the nitrogen status of the soil, each crop removing 45 to 60 kg ha^{-1} N. Work in the Central Plain of Thailand suggests that a 3 t ha^{-1} rice crop is virtually independent for nitrogen since one, two, or three crops a year can be grown without added nitrogen and yields are approximately the same for each crop (Walcott et al., 1977).

Nonsymbiotic nitrogen-fixing bacteria are known to be present in tropical forest and savannah soils, and it is generally believed that part of the accumulation of nitrogen in the biomass of natural vegetation during the recovery phase of shifting cultivation systems can be attributed to this source (Nye and Greenland, 1960). Furthermore, Dobereiner and her colleagues in Brazil have demonstrated nitrogen fixation in the rhizosphere of tropical grasses (Dobereiner et al., 1972; Day et al., 1975), and there is evidence of a symbiotic nitrogen-fixing association in some grass species (Dobereiner and Day, 1974). The significance of these organisms in the soil nitrogen balance has yet to be assessed.

In the past few years, evidence has been accumulating for the uptake by crop plants of ammonia from the atmosphere via stomata (Hutchinson et al., 1972; Wetselaar et al., 1977). On an individual farm scale the quantities involved are small: for example, Wetselaar et al. concluded that a flooded rice crop absorbed about 2 kg ha^{-1} ammonia nitrogen from the air during a two- to three-week period after ammonium sulphate fertilizer had been applied to the soil and had partially volatilized.

■ 4.2.5 *Nutrients in rainfall and run-on water.* In agricultural regions close to industrial areas there may be a significant return of nitrogen and sulfur to the soil via rainfall, the source being nitrogen and sulfur compounds released into the atmosphere from factory processes. However, the nutrient gain from rainfall in nonindustrial areas is a disputed question, both in respect of its magnitude and the source of the nutrients (Stevenson, 1965). There is good evidence to suggest that it is largely a territorial redistribution of soil dust, blown into the atmosphere at one site and washed out of it at another (Wetselaar and Hutton, 1963). In any case the amounts involved in nonindustrial localities are fairly low.

Nutrients in soil and organic matter that are suspended in run-on water, that is, eroded and carried from elsewhere, may be a significant input in certain situations: for example, in rice-growing areas subject to inundation from silt-laden rivers or in riverine cropping systems that involve planting on previously inundated land when the seasonal river flow declines.

■ 4.2.6 *Nutrients brought in.* In tropical farming systems an input of nutrients via stockfeed brought in, subsequently entering the crop nutrient cycle through animal residues, is uncommon since livestock production based on purchased feed is rare. One important exception is intensive vegetable-growing around Southeast Asian cities, which is often associated with pig and poultry raising. Although the livestock subsist partly on vegetable residues, there is normally a purchased input of feed protein and the manure is utilized on cropfields.

■ 4.2.7 *Fertilizers.* In high-technology crop farming the nutrient input-output balance is frequently dominated by the fertilizer component, but in tropical cropping systems in general, purchased fertilizer inputs are low. In 1969, only about 12 percent of the fertilizer used in the world was consumed in the tropics, although nearly half the world's land area is tropical. The important exceptions to low fertilizer use are in plantation perennial crop systems, in intensive irrigated vegetable agriculture, and in the high input–high output systems of irrigated wheat and rice growing that have developed as a result of research at CIMMYT in Mexico and IRRI in the Philippines. Englestad and Russel (1975) give data on the rates of increase in production and consumption of fertilizer nitrogen, phosphorus, and potassium in the tropics.

The major limitations to the use of fertilizer in the tropics are the variability in yield response under rainfed conditions; poor extension services and hence a lack of appreciation on the part of the farming community of the potential value of fertilizer; poor infrastructure in respect of transport, storage, and distribution of fertilizer; and limited farm credit facilities.

The technical limitations to the efficiency of applied fertilizers in the tropics are those considered in this chapter in relation to nutrients in general: leaching of nitrogen and potassium, denitrification under flooded conditions, volatilization of ammonia (particularly serious with surface-applied urea), and fixation of phosphorus.

Although tropical crop farmers lack the advanced fertilizer application equipment of western crop agriculture, the fact that planting operations are by hand gives flexibility. For example, fertilizers may be readily placed in any position with a dibble stick. The "mudball" technique for placement of fertilizer for rice is an excellent example of simple technology for tropical farmers (IRRI, 1976).

■ 4.3 THE BIOGEOCHEMICAL CYCLE: OUTPUTS

■ 4.3.1 The nutrient outputs may be classified as:

a. Nutrients in crop material that leaves the farm.
b. Nutrients in stock or stock products that leave the farm.
c. Leaching of nutrients below the root zone.
d. Losses of nitrogen to the atmosphere through volatilization or denitrification.
e. Losses of nitrogen and sulfur to the atmosphere through burning of vegetation or crop residues.
f. Losses through runoff water, i.e., soil erosion.

■ 4.3.2 *Crops leaving the farm.* Although it is not uncommon to see figures for the average amount of minerals removed by crops, such values are highly dependent upon soil fertility levels. Removal of nutrients per unit yield of crop provides a rather better index. These values are reasonably reliable for the grain yield of seed-bearing crops, since the nutrient content of mature seeds tends to be moderately stable within a range of fertility levels, but they are less dependable for tubers.

The ratio of nutrients in that part of the crop which is removed to that part which is returned to the soil or consumed by stock is an important feature of the crop nutrient balance. The proportions by weight and by nutrient yield of the economic yield of tops and roots vary widely not only between crops but may also vary between cultivars of the same crop. Thus the proportion of total aboveground nitrogen found in the grain of an improved sorghum cultivar with a high harvest index might be 60 percent; for an unimproved local cultivar total aboveground nitrogen might be the same, but harvest index would be lower and the proportion of grain nitrogen might not exceed 25 percent. Wetselaar (1967b) found very large differences in the relative proportions of nitrogen in crop legumes removed and returned to the field according to the manner of handling the crop.

Lastly, though we normally think of crop export from the farm in terms of material that will be consumed or utilized by man, in India crop residues, e.g., sorghum stover, are frequently sold off the farm for cattle or buffalo employed in road transport or in urban dairies.

■ 4.3.3 *Stock products leaving the farm.* In intensive systems of animal production that involve the daily output of high-nutrient stock products such as milk and eggs, this component of the biogeochemical balance assumes some importance. However, intensive egg production is not a major tropical farming system and milk production for sale from crop farms is only of importance on the Indian subcontinent. Furthermore, the amounts per farm per day are usually small. If meat animals are a sale product of the farm, this is scarcely a significant nutrient loss since the retention of nutrients in stock tissues is relatively low and a high proportion is voided as dung and urine. Hence this output component is not of major biogeochemical significance in tropical farming systems.

■ 4.3.4 *Leaching.* Of the main nutrient elements leaching is normally only of major importance with respect to nitrogen and to a lesser degree potassium. Leaching of phosphate may occur on coarse-textured sands and organic soils, but it is not common (Engelstad and Russel, 1975). The degree of leaching will be dependent on the drainage characteristics of the soil and on the amount of water passing through the crop root zone.

A good relation has been demonstrated on a free-draining soil in a north Australian summer-rainfall climate between total annual rainfall and the mean depth of accumulation of leached nitrate (Wetselaar, 1962). In heavier soils with poor drainage, nitrate nitrogen may move out of the topsoil but is unlikely to be leached to any depth and may be available for uptake later by crops as their root systems develop. On the other hand, poor drainage may restrict the depth of rooting of crops and hence their subsequent uptake of leached nitrate. It must be admitted that we know little about the movement and uptake of nitrogen under conditions of poor drainage.

On free-draining soils nitrate nitrogen can move well below the root zone of some annual crops. Hence the rooting depth of the crop may determine whether or not leached nitrogen is recovered. For example, in North Australia a grain sorghum crop significantly depleted soil nitrate nitrogen only to a depth of 60 cm, whereas pearl millet took up nitrate to a depth of 150 cm (Wetselaar and Norman, 1960).

Leaching effects are particularly important in early wet season in summer rainfall climates. The main flush of mineralization of soil organic nitrogen occurs then, and if on free-draining soils crop sowing is delayed a proportion of the nitrate released may be leached before the roots of the developing crop can intercept it.

Under perennial crops with permanent root systems leaching is of minor significance, since any nitrate leached from the surface is likely to be recovered by deeper roots. It has been suggested that the high rate of nutrient accumulation by recovering vegetation in the early fallow phase of shifting cultivation systems may be partially attributable to the recovery of nutrients leached out of the root zone of annuals grown in the preceding cropping break.

■ 4.3.5 *Volatilization and denitrification.* Nitrogen may be lost out of the farm system to the atmosphere through denitrification to elemental nitrogen under anaerobic conditions: typical situations are the temporary flooding of upland crops or where fertilizer nitrogen is surface-applied to flooded rice. Losses of nitrogen to the atmosphere as volatilized ammonia are favored by a wet surface soil, high temperature, and alkalinity. Volatilization from urine deposits may be a significant source of loss in the cycle of nutrient return from animal residues.

■ 4.3.6 *Burning.* There are two categories of plant material that may be burned in tropical farming systems: crop residues and natural vegetation. When plant material is burned, nitrogen and sulfur are lost to the atmosphere as gaseous oxides, but other nutrients are retained in the ash. Burning of natural vegetation is, of course, a key element of the biogeochemical pattern in all cropping systems involving a rest or fallow phase and will be treated in more detail in later chapters.

The fate of crop residues in tropical farming is various: they may be allowed to decay, be consumed *in situ* by stock, be harvested and removed either for consumption by stock or for other purposes (e.g., roofing materials, fuel, etc.), or they may be burned *in situ* or elsewhere. Burning of rice straw is common in Southeast Asia: although this results in a loss of nutrients from the system there is little evidence to suggest any significant effect on subsequent yield compared with no burning, perhaps because of the resilience of the flooded-rice system by virtue of its nitrogen-fixing capacity.

■ 4.3.7 *Runoff.* If topsoil is lost from the farm through erosion this represents a significant drain of nutrients, particularly since soil nutrient concentrations are normally higher near the surface than at depth. In addition, the nutrient concentration of soil in runoff water tends to be higher than that of the soil from whence it came since the organic fraction is selectively removed. This *enrichment ratio*, which applies to the elements closely associated with the organic fraction (that is, nitrogen, sulfur, and to a lesser extent phosphorus, but not to potassium, calcium or magnesium), is commonly between 2 and 3.

■ 4.4 Internal biogeochemical cycles

■ 4.4.1 In the two foregoing sections sources of nutrient gain or loss to the farm system have been considered. In addition, within the boundaries of the farm itself a series of internal nutrient cycling processes is operating. The more important of these are (a) uptake of soil nutrients by crops; (b) the return of crop residues to the soil; (c) food consumption and excretion by humans; (d) feed consumption and excretion by stock; (e) consumption of stock or stock products by humans.

■ 4.4.2 *Crop nutrient uptake.* The uptake of available soil nutrients by crops is, of course, the basic starting point of the biogeochemical cycle in cropping systems, if cycles may be said to have starting points. Its pattern and magnitude are dependent on a range of conditions: the concentration of soluble nutrients at different levels in the crop root zone; the movement of ions in the soil; the root development pattern of the crop in time and space; the activity of the crop rhizosphere; the availability of soil water in time and space in the crop root zone, and so forth.

■ 4.4.3 *Crop residues.* As indicated in section 4.3.6, the end-use of crop residues varies, though the choices open to the individual farmer may be rather narrow. He may be forced to save residues to feed draft animals or even to sell residues for income. Handharvesting processes may involve removal of the whole plant from the field, as with pigeon pea, the residue of which is fit only for burning. If the residues are woody, e.g., cassava stems and cotton bushes, and if heavy equipment to chop them into the soil is not available to him, he has little alternative but to burn. The composting of crop residues, with or without animal manure, with the planned aim of returning them to cropfields, is comparatively rare and generally confined to intensive vegetable-growing systems. Except in such systems, where high inputs are economic, or where extreme shortage of land has forced farmers into intensive nutrient-conserving procedures, it is broadly true to say that the small-scale tropical crop farmer does not, and indeed cannot afford to, think in terms of the long-term nutrient status of his cropland; his decisions on secondary matters such as crop residue disposal are governed by short-term considerations. If he does not need to conserve them for stock or other domestic purposes, if they do not decay naturally, or if they cannot be grazed off with a low labor input, he will probably burn them as the easiest way out.

Mention should be made here of green manure crops, which often figure prominently in experimental programs and in tropical agricultural science literature. In practice, the growing of legume crops specifically for incorporation into the soil is of very limited occurrence in tropical farming systems: the intensive irrigated cropping of subtropical Taiwan is an instance of this practice.

■ 4.4.4 *The human food cycle.* The proportion of total farm food output consumed by the farm family is, of course, a direct measure of the subsistence element in the farm system. In terms of nutrient cycling, the interest lies in the extent to which the nutrients consumed are returned to the system either as household waste (e.g., animal viscera and bones, vegetable material) or as excreta.

In western farming, with established systems of garbage and sewage disposal, this cycle is almost nonexistent, but it assumes some significance in less-developed regions. In irrigation areas, supply and drainage canals may be used as toilets, which implies a certain circulation of nutrients within a region even if they are lost to the specific farm. In sedentary rainfed farming systems a concentric pattern of fertility tends to develop around domiciles and villages. This may be partially the result of the distribution of animal manure which, because of its bulk, is more likely to be applied to cropfields and gardens near the domicile than to those further out. However, the pattern may also represent a steady accretion of household waste and human excreta. In parts of Africa, domiciles may be moved at intervals not merely because it is often easier to build a new house than to repair an old one, but also in order that the accumulated fertility of the old site may be realized by cultivating it for crops.

■ 4.4.5 *The stock feed cycle.* Livestock in tropical crop farming systems are normally housed, at least at night, close to the domicile. This leads in time to an accumulation of nutrients in the form of animal residues at a central focus on the farm, the original nutrient source being, of course, dependent on the diet of the stock. For ruminant stock, if a high proportion of their feed is from crop residues, this represents a cycling from cropfields. If a high proportion of the feed is forage from waste land, then the stock are effectively concentrating nutrients that would otherwise merely form part of the local ecological cycles of growth and decay. For nonruminants such as pigs or poultry, the original source of nutrients may be the farm fields, when homegrown grains or tubers are fed, or may be outside the farm when feed is brought in.

Systematic return of animal residues to cropfields is only practiced in some areas of the tropics. In many instances the practice has gradually been forced upon farmers by increasing population density and declining soil fertility; a substitute for the fallow phase of less intensive cropping systems for which the land area is no longer adequate. There is a far larger range of instances of comparative unsystematic return of residues. The dung that accumulates near the domicile has to be disposed of periodically, and there is a tendency to apply it to vegetable and fruit tree gardens around the domicile and to immediately adjacent cropfields.

Crop residues may also be recycled through the animal by grazing *in situ*, and in some instances this is controlled to the extent of fencing or penning stock on a specific cropfield. Another low-input method of recycling is met with in crop/cattle systems in South America, where the subsistence food garden is placed downslope of the stockyard and receives nutrients in run-on water.

In areas of dense human and ruminant populations and shortage of natural fuel, cattle and buffalo dung are used for domestic heating and cooking. A simple hypothetical but realistic example illustrates the significance of this nutrient drain. Assume that the total aboveground nitrogen yield of an unfertilized sorghum crop in semi-arid India is 40 kg ha^{-1} N, derived wholly from mineralized soil nitrogen. Assuming 15 kg ha^{-1} N in the grain, this leaves 25 kg ha^{-1} N in the residue, which is wholly consumed by stock. If the animals retain 15 percent of this in tissue, 21 kg ha^{-1} N is voided. Assuming one-third of this to be in the dung, which is collected and burned, this represents a loss of 7 kg ha^{-1} N to the atmosphere, equivalent to about half that taken off in grain.

■ 4.4.6 *Consumption of stock by humans.* The importance of this pathway in internal nutrient cycling varies greatly with the farming system. Generally speaking, in tropical crop farming only nonruminants are raised specifically for home consumption. Large ruminants, cattle or buffalo, are normally only eaten by farm families on special occasions: following death of or injury to the animals or to fulfill cultural obligations—ceremonies, formal hospitality, sacrifices, etc. These limitations apply even to pastoral systems, though in this instance there is heavy dependence on ruminant products, i.e., milk and cheese. The status of small ruminants, sheep and goats, as a locally consumed food is intermediate between that of nonruminants and large ruminants.

REFERENCES

Day, J. M., Neves, M. C. P., and Dobereiner, J. 1975. Nitrogen fixation on the roots of tropical grasses. *Soil Biology and Biochemistry* 7:107.

Dobereiner, J., and Day, J. M. 1974. *Associative symbiosis in tropical grasses. Characterization of microorganisms and di-nitrogen fixing sites.* Pullman, Washington: International Symposium on Nitrogen Fixation.

Dobereiner, J., Day, J. M., and Dart, P. J. 1972. Nitrogenous activity and oxygen sensitivity of the *Paspalum notatum—Azotobacter paspali* association. *Journal of General Microbiology* 71:103.

Engelstad, O. P., and Russel, D. A. 1975. Fertilizers for use under tropical conditions. *Advances in Agronomy* 27:175.

Eyre, S. R. 1963. *Vegetation and soils. A world picture.* Chicago: Aldine.

Firth, P., Thitipoca, H., Suthipradit, S., Wetselaar, R., and Beech, D. F. 1973. Nitrogen balance studies in the central plain of Thailand. *Soil Biology and Biochemistry* 5:41.

Hutchinson, G. L., Millington, R. J., and Peters, D. B. 1972. Atmospheric ammonia: Absorption by plant leaves. *Science* 175:772.

International Rice Research Institute. 1976. *The* IRRI *Reporter,* no. 2/76. Los Banos, Philippines.

Kamprath, E. J. 1973. Phosphorus. In: Sanchez, P. A., ed. *A review of soils research in tropical Latin America.* North Carolina Agricultural Experiment Station Technical Bulletin no. 29.

Moss, R. P. 1969. The ecological background to land use studies in tropical Africa, with special reference to the West. In: Thomas, M. F., and Whittington, G. W., eds. *Environment and land use in Africa.* London: Methuen.

Nye, P. H., and Greenland, D. J. 1960. *The soil under shifting cultivation.* Harpenden, England: Commonwealth Bureau of Soils, Technical Communication no. 51.

Patrick, W. H., and Mahapatra, I. C. 1968. Transformation and availability to rice of nitrogen and phosphorus in waterlogged soils. *Advances in Agronomy* 20:323.

Stevenson, F. J. 1965. Origin and distribution of nitrogen in the soil. In: Bartholomew, W. R., and Clark, F. E., eds. *Soil nitrogen.* Madison, Wisconsin: American Society of Agronomy.

Walcott, J. J., Chauviroj, M., Chinchest, A., Choticheuy, P., Ferraris, R., and Norman, B. W. 1977. The long-term productivity of intensive rice-cropping systems on the Central Plain of Thailand. *Experimental Agriculture* 13:305.

Watanabe, I., Lee, K. K., Alimagno, B. V., Sato, M., del Rosario, D. C., and de Guzman, M. R. 1977. Biological nitrogen fixation in paddy field studied by *in situ* acetylene-reduction assays. IRRI Research Paper Series no. 3.

Wetselaar, R. 1962. Nitrate distribution in tropical soils. III. Downward movement and accumulation of nitrate in the sub-soil. *Plant and Soil* 16:19.

Wetselaar, R. 1967a. Determination of the mineralization coefficient of soil organic nitrogen on two soils at Katherine, N.T. *Australian Journal of Experimental Agriculture and Animal Husbandry* 7:266.

Wetselaar, R. 1967b. Estimation of nitrogen fixation by four legumes in a dry monsoon area of north-western Australia. *Australian Journal of Experimental Agriculture and Animal Husbandry* 7:518.

Wetselaar, R., and Hutton, J. T. 1963. The ionic composition of rainwater at Katherine, N.T., and its part in the cycling of plant nutrients. *Australian Journal of Agricultural Research* 14:319.

Wetselaar, R., and Norman, M. J. T. 1960. Recovery of available soil nitrogen by annual fodder crops at Katherine, N.T. *Australian Journal of Agricultural Research* 11:693.

Wetselaar, R., Shaw, T., Firth, P., Oupathoum, J., and Thitipoca, H. 1977. Ammonia volatilization losses from variously placed ammonium sulphate under lowland rice field conditions in Central Thailand. Seminar on Soil Environment and Fertility Management in Intensive Agriculture. Tokyo. October 1977.

Yoshida, T., and Ancajas, R. R. 1973. Nitrogen fixing activity in upland and flooded rice fields. *Soil Science Society of America Proceedings* 37:42.

Chapter 5

General Socioeconomic Background

■ 5.1 Introduction

This book is concerned with the biological and physical processes operating in tropical farming systems and does not attempt to treat economic issues in any detail. There have been numerous studies of the microeconomics and sociology of tropical farms and farming communities, and the principles emerging from these have been embodied in a number of useful texts. The outstanding example is Ruthenberg (1971); other invaluable studies include Boserup (1965), Wharton (1969), Clark and Haswell (1970), and Haswell (1973). However, the structure of tropical farming is in many ways so different from that of western high-technology agriculture that a brief review of its most important socioeconomic characteristics is appropriate. In this chapter particular emphasis is placed on subsistence cropping, since the high proportion of subsistence production units in tropical farming is one of its main distinctions from advanced temperate-zone agriculture.

■ 5.2 Subsistence farming: definition and characteristics

■ 5.2.1 The word "subsistence" is in common use in relation to tropical agriculture with two quite distinct meanings: it is used with reference to dietary standards to signify a barely adequate food intake, and is used in the socioeconomic sense with reference to the proportion of goods and services produced and consumed on the farm or purchased and disposed of elsewhere. We are here concerned with the latter usage.

Pure subsistence, as defined by Wharton (1969), "refers to a self-contained unit where all production is consumed and none is sold and where no consumer or producer goods and services from sources external to the unit are purchased." Such a situation scarcely exists today: Though there are many farm units that do not purchase food, virtually all of necessity trade some of their products to permit at the very minimum the purchase of salt, axes, hoes, kerosene, clothes, transistor radios, etc. More realistically, we can apply the term subsistence farming in an imprecise way to farm systems in which the main thrust of the production effort is directed toward supplying the food requirements of the production unit itself.

Wharton summarizes the socioeconomic features which characterize, though they do not define exclusively, subsistence farming:

a. The ratio of consumption to sale of farm products is high.
b. The ratio of hired labor input to total labor input is low.
c. The ratio of purchased factor inputs to total factor inputs is low.
d. The level of production technology is low.
e. The level of income is low.
f. The socioeconomic restrictions on decision-making are substantial.
g. The influence of noneconomic factors on decision-making is substantial.
h. The influence of interpersonal relations in the pattern of activity is substantial.

■ 5.2.2 Characteristic (g) above, the influence of noneconomic factors on decision-making, brings into focus the much-argued question: To what degree can the subsistence farmer be regarded as a typical representative of "economic man"? Penny (1969) uses the rather cumbersome term "economic-mindedness" to denote the extent to which the farmer is motivated towards improving his economic status. His aphorism "all farmers make economic choices, but only some of them make choices that are economic" encapsulates the situation, and one of the prime tasks in agricultural development is to create conditions under which it is possible for the farmer to make choices that are economic. The weight of evidence, from the numerous studies made in postwar years of subsistence farmer attitudes, indicates that when such conditions are created cultural barriers to the acceptance of changes in technology are not a major impediment.

■ 5.3 Forms of economic organization

■ 5.3.1 In chapter 1, the taxonomy of tropical farming systems was briefly outlined, and it was pointed out that any all-embracing classification required a complex of distinguishing criteria: cultivation frequency (e.g., shifting cultivation/intensive cultivation), type of product (crop farms/livestock enterprises), use of water (rainfed systems/irrigated systems) and economic organization. However, it is possible to categorize farming systems in a rather simplistic manner on the basis of economic organization alone, the criteria being rights to land and managerial structure; thus:

a. Shifting cultivation and nomadic or seminomadic pastoral systems.
b. "Permanent" farming systems of simple managerial structure.
c. Plantation systems.

■ 5.3.2 Shifting cultivation and seminomadic or nomadic pastoral systems may seem at first sight strange bedfellows, but there is a key socioeconomic distinction between these two systems and all others in respect of the definition of the land area being utilized and user's rights to that land. Within this group the operators and their families constitute managerially simple and economically independent small units, a characteristic shared with group (b) and a distinction from group (c); but in no case do they restrict themselves to a cadastrally definable land area for their agricultural operations. Almost universally, customary rights to the use of an approximately demarcated land area have developed over a period of time, rights that may function at more than one level of social organization (an ethnic group, a tribe, a village, a family), but legal titles do not exist and land is not a marketable commodity.

■ 5.3.3 The second group in this categorization is described as "permanent." The word is in inverted commas to emphasize the distinction between permanence in relation to farm boundaries and permanence as used, for example, by Ruthenberg (1971) to signify a cropping system of high cultivation frequency. The key criterion which distinguishes group (b) from group (a) is that the farm unit operates within a defined land area, to which the owner or tenant has some rights, if at most only the right to cultivate it for a single season and to take some proportion of the crop for his effort. In socially well-organized countries, the land area in question will be defined on a map held by a public authority. Where legal titles exist, and sometimes where they do not, land may be bought, sold, or leased. When there are no titles, a distinction between groups (a) and (b) still exists in that in (b) custom has determined the *precise* boundaries of the holding. Such characteristics apply to farm units devoted either to crop or livestock production. The definition of group (b) includes also the phrase "of simple managerial structure," which serves to distinguish it from group (c).

■ 5.3.4 Plantation agricultural systems are distinguished from groups (a) and (b) by their characteristic patterns of labor organization and managerial structure, which approach those of an industrial unit. Plantations may be individually owned or company owned, but there is typically a manager with technical if limited economic control over operations, directing a stratified labor force, with foremen and labor groups allocated specific production tasks: e.g., the tappers and factory operators on a rubber plantation. The distinction between plantations and other forms of organization is not wholly clear-cut, however, since there are intermediate types between the typical large plantation organization as described above and small holdings devoted to perennial "plantation" crops. These have a much simplified labor and managerial structure and would be classified within group (b).

This book is not concerned with plantation systems, but the above point has been made to stress that though to an increasing degree Western farming units are developing along more complex organizational lines, such patterns scarcely exist in tropical farming outside the traditional plantation system. Furthermore, the peak of classical plantation agriculture was reached in the first half of this century; associated with the post-war dissolution of colonial empires is a trend toward smaller and simpler units.

■ 5.4 Man-land ratios

■ 5.4.1 The pressure of agricultural man upon land resources may be expressed by three measures, in increasing order of utility and precision: first, normal population density, that is, total population per unit area; second, nutritional or physiological density, or total population per unit area of cropland; third, agricultural population per unit area or, as conventionally expressed, land area per adult male engaged in agriculture. According to Haswell (1973), in low-income tropical countries the adult males engaged in agriculture average about one quarter of the total rural population.

It is clearly difficult to generalize about the average land requirement of various types of farming systems, but this has not prevented economists and sociologists from making the attempt in respect of subsistence farming. Haswell (1973) has reviewed these efforts. Estimates range from 250 to 400 kg grain equivalent per person or, in energy terms, about 4,000 to 6,000 MJ for the amount of energy food regarded as a minimum annual production target to ensure an adequate diet for subsistence cultivators.

The land area required to sustain any given level of production will naturally depend on the primary resource characteristics of the land and the level of technology available to exploit the resource. A good example of a careful attempt to assign capability ratings to land regions on the basis of traditional agricultural systems is that of Allan (1965), who formulated a whole series of capability classes for regions of Africa on the criterion of their capacity to sustain cropping systems of varying cultivation frequency. Allan found fairly consistent values for crop area per head of rural population for the main climatic zones. Thus in savannah climates, where the main energy crops were cereals, values ranged from 0.4 to 0.7 ha $head^{-1}$, while in twin-peak rainfall or equatorial climates where the main energy crops were tubers and bananas, the range was 0.2 to 0.4 ha $head^{-1}$.

Haswell (1973) suggests that with no limitation on land, subsistence farm size will expand up to about 2 ha of cropland per adult agricultural male and somewhat less in areas of difficult topography. (The upper limit could be imposed either by the area that can be handled with the family labor force or merely by the adequacy of the area to provide subsistence, as in Allan's equatorial class of cultivators.) This would be equivalent to about 0.5 ha per head of rural population. The possession of draft animals will, of course, allow a larger area to be cultivated. In general, Charreau (1974) has estimated the upper limits for farms with a pair of oxen to lie between 8 and 16 ha.

■ 5.4.2 Given the appropriate climate and soil resources and a technology that permits more intensive yet stable exploitation, the area farmed per adult agricultural male may fall to less than 1 ha, or 0.2 ha per head of rural population, as in many Asian rice-growing areas.

Some idea of the level of population density that can be supported on intensively cultivated fertile land is given in Table 5.1, which is based on data from Grigg (1974) for Java in the past century and a half. It is believed that in the early nineteenth century wet rice cultivation (*sawah*) was confined to inland valleys, and was complemented by shifting cultivation (*ladang*) in the hills. A major expansion in wet rice area in the lowlands took place in the late nineteenth century, and in the twentieth century intensive rainfed cropping (*tegalan*) has increased greatly, often in areas of sharp relief. Shifting cultivation is now a residual farming system in Java. The capacity of wet rice land to respond in productivity to increasing inputs of careful husbandry even within the context of traditional technology has been stressed by Geertz (1963).

It will be noted from Table 5.1 that crop area per head was already below 0.2 ha in 1940. The current population of Java exceeds 80 million. It may be calculated that a nutritional density of, say, 0.12 ha per

TABLE 5.1
Population and cropland in Java and Madura, 1815–1960

Year	Cropland (km^2)	Population ($\times 10^6$)	Nutritional Density (km^{-2})	Cropland per head (ha)
1815	15,150	5.0	330	0.30
1900	66,660	28.4	426	0.23
1920	80,800	34.4	425	0.23
1940	90,500	48.0	530	0.18
1960	88,880	63.0	708	0.14

SOURCE: Grigg (1974).

head and an average yield in grain equivalents of 2 t ha^{-1} implies a production level of 240 kg $head^{-1}\ yr^{-1}$, close to the minimum requirement for subsistence.

■ 5.4.3 Where there is pressure on land, irrigated farms tend to be smaller than rainfed farms in the same environment. Table 5.2 from Kampen (1974) illustrates the relation between farm size and irrigation in two states of India. On average, the percentage of farms in Andra Pradesh of less than 1 ha is more than twice as high in districts with substantial irrigation development than in districts of predominantly rainfed cropping. In Karnataka the percentage is more than three times as high. Within the predominantly rainfed districts of Andra Pradesh, the average size of wholly rainfed farms is 1.9 to 3.3 ha and of wholly irrigated farms 0.5 to 0.9 ha.

■ 5.5 LANDOWNERSHIP AND TENURE

■ 5.5.1 It is virtually impossible to generalize safely about the conditions of landownership and tenure in tropical farming systems. Attention has already been drawn to the special cases of shifting cultivation and nomadic pastoral husbandry, where land rights are established and maintained by custom only over imprecisely defined areas; but within the category of small permanent farming systems there is a vast range of ownership and tenure patterns. Historical factors are often a strong determinant: for example, in the contrast between the high proportion of owner-operators in Thailand and the low proportion in the Philippines. Within any one country there may be regional variations related to ecological factors: for instance, the contrast between the small owner-operators of the upper Central Plain of Thailand, where water control is feasible and transplanted rice is largely grown, and the much larger landlord-tenant farms of the lower Central Plain, where water control is poor and broadcast rice is grown with mechanized land preparation. The range of tenancy and sharecropping contracts is equally wide, from registered long-term whole-farm tenancies down to oral agreements to sharecrop a single field for a single cropping period. A range of communal landholding systems still survives in Africa.

TABLE 5.2
Farm size distribution in predominantly rainfed and predominantly irrigated districts in two Indian states

State and category	District	Farm size classes (ha)						
		<1	1–2	2–3	3–4	4–5	5–10	>10
Andra Pradesh Rainfed	Adilabad	26	17	19	9	7	15	6
	Mahboobnagar	21	20	16	10	8	17	1
	Medak	43	22	11	6	4	8	5
Andra Pradesh Irrigated	E. Godovari	70	16	3	3	2	3	1
	Krishna	64	19	7	4	2	3	1
	W. Godovari	62	19	8	4	2	4	1
Karnataka Rainfed	Bellary	16	21	17	11	9	17	9
	Bijapur	10	16	14	11	8	24	17
	Gulbarga	11	15	13	11	9	24	17
Karnataka Irrigated	Mandya	55	24	10	5	2	3	—
	Shimoga	33	30	15	7	5	7	2
	S. Kanara	47	28	12	5	3	4	1

SOURCE: Kampen (1974).

■ 5.5.2 *Fragmentation.* One feature of landownership characteristic of closely settled regions of the tropics is the fragmentation of individual holdings into physically separate field units, which may be scattered over an appreciable geographical area. Table 5.3 demonstrates this for semi-arid India (approximately those areas in rainfall zones of 500 to 1,500 mm). Even holdings as small as 0.2 to 1.0 ha are likely to be fragmented into three separate parcels, and farms of 20 ha average ten separate units of land. It is also worth noting that in semi-arid India over one-third of the holdings are of less than 0.2 ha.

TABLE 5.3
Fragmentation of farms in semi-arid India

Size class of holding (ha)	No. of land parcels	Mean area of parcels (ha)	% of holdings of area less than upper class limit
0 – 0.19	1.63	0.06	34.6
0.20 – 0.39	2.70	0.16	39.4
0.40 – 0.99	3.14	0.22	54.1
1.00 – 1.99	4.50	0.34	69.3
2.00 – 2.99	5.10	0.51	78.8
3.00 – 3.99	6.01	0.63	84.3
4.00 – 4.99	6.02	0.78	88.3
5.00 – 5.99	6.76	0.91	90.7
6.00 – 7.99	7.18	1.07	94.0
8.00 – 9.99	8.27	1.20	96.0
10.00 – 11.99	8.12	1.48	97.1
12.00 – 19.99	9.65	1.92	99.1
20.00 +	11.52	3.09	100.0

SOURCE: Ryan (1974).

Haswell (1973) points out that degree of fragmentation is related to security of title. If land has a clear title, it is possible to buy and sell uneconomic parcels and thus to mitigate in some measure the fissile effects of inheritance. If titles are uncertain, land parcels are less likely to be traded since ownership cannot be clearly established.

Land values in permanent tropical farming systems are, as elsewhere, determined by the inherent productivity of the land, the general pressure of demand related to population density, the degree of improvement for cropping, and access to sites of product disposal and supply. Haswell's (1973) data (Table 5.4) from Mindanao in the South Philippines illustrates the progressive increase in land values with level of improvement, which determines the permissible level of intensification of the farming system.

TABLE 5.4
Land values and level of land improvement. Philippines

		Land value in paddy rice equivalents (t ha^{-1})
a.	Land suitable for shifting cultivation (return from secondary forest trees).	0.09
b.	Partially cleared land, suitable for more intensive hoe cultivation.	0.26
c.	Fully cleared land, suitable for cultivation with draft animals.	2.12
d.	Fully cleared land laid out for gravity flow irrigation.	3.53

■ 5.6 Labor Input

■ 5.6.1 The input of labor in tropical farming systems may be viewed from three aspects: input per unit time, per unit area, and per unit of crop production.

■ 5.6.2 *Input per unit time* is normally expressed as total number of working hours per year or as average working hours per day. The average hours worked per day in tropical cropping systems is often surprisingly low.

Boserup (1965), reviewing data on shifting cultivation systems from a number of sources, concluded that "those investigations of which I am aware agree in reporting a very small average labor input per man and year in agriculture under the system of forest fallow." Later it will be seen that this may not be a valid generalization for shifting cultivation. The estimation of total labor in shifting cultivation systems is made difficult by the fact that food procurement often includes a substantial hunting or foodgathering component. Furthermore, the proportion of time used in walking to and from scattered cropfields tends to be high. It should also be kept in mind that in those more primitive societies where food procurement is relatively easy the human states of "working" and "not working" are not so polarized as in modern society. Hunting, for example, though perhaps a necessity, can also be a pleasurable occupation.

In relation to semi-intensive and intensive rainfed cropping systems, Norman (1968), basing his estimates on West African data, states that "it would appear realistic to expect male adults to work not much more than 5.0 hours a working day and a total of 1,200 hours or about 230 days a year." Haswell (1973) quotes data of her own from Gambia, where adult males worked 5.7 hours a day and 110 days a year, a total of 627 hours. Ruthenberg (1971), using information from three semi-intensive cropping systems, gives figures of 709, 857, and 1,014 hours a year for adult males.

The above values are for hoe cultivation systems. The introduction of draft animals may reduce human labor input per hectare of cropland, but the degree to which their maintenance is likely to increase total labor input per year depends greatly on the amount of effort required to feed them. Boserup (1965) quotes data from India indicating that where the bulk of draft animal feed comes from grazing, animal care labor may be less than 10 percent of the total labor input for crop production, whereas when animals are maintained largely on crop products and residues the figure may be 50 percent.

Irrigated cropping systems normally demand a higher total input of labor per year, though, as will be seen below, this may be due largely to an extension through irrigation of the period over which crop production is possible: that is, an increase in the number of working days and not necessarily a longer working day. Important additional inputs associated with irrigation are the maintenance of supply channels and drains and, particularly when water has to be raised from streams or wells by human or animal power, the actual labor of bringing water to the crops.

■ 5.6.3 *Seasonal distribution of labor.* As noted earlier, a low average work input over the year is often the result of a limited cropping season. Clearly, therefore, seasonal labor distribution is in many instances a function of climate: the no-crop period of enforced idleness may vary from nearly zero in a climate where cropping is possible all the year to perhaps 7 to 8 months in a short-season summer-rainfall environment. In addition, however, there is a general relation between seasonality of labor demand and type of farming system.

The most highly seasonal labor distribution patterns are found for intensive and semi-intensive rainfed cropping systems without draft animals in short-season climates. With draft animals, the seasonal pattern of labor for crop production may differ little, but the year-round care and maintenance of stock, particularly if they subsist to any great extent on crop products and residues, tend to dampen the contrast between the cropping and no-cropping periods. Shifting cultivation systems normally exhibit a less extreme seasonality than intensive rainfed systems, for the rainy season tasks of crop production are balanced by the dry season tasks of fallow clearing and burning.

Where irrigation permits an extension of the cropping season beyond the normal rainy period, seasonal distribution of labor is much improved. In general the more intensive the cropping system the more even the distribution throughout the year and the higher the total labor input per unit area. This is illustrated in Figure 5.1, which compares one-

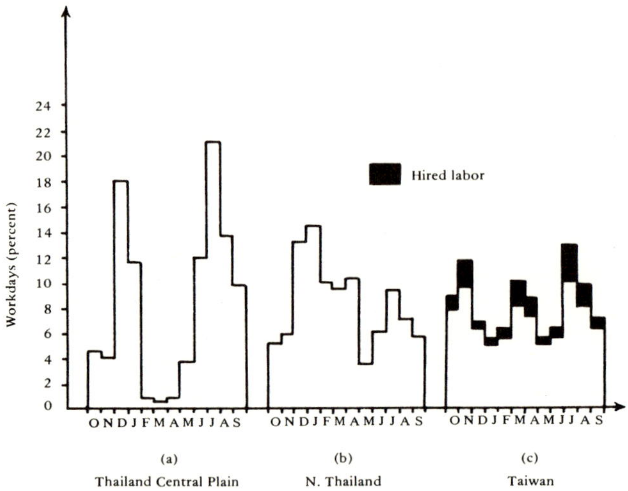

Figure 5.1. Seasonal distribution of labor inputs with increasing intensity of rice-based cropping (Ruthenberg, 1971).

crop rainfed wet rice in Central Thailand, a two-crop system of irrigated rice and upland crops in North Thailand, and intensive irrigated cropping in Taiwan, where two rice crops and at least one upland crop are grown in the year.

■ 5.6.4 *Input per unit area.* Table 5.5 summarizes a series of estimates of labor input per hectare for shifting cultivation, semi-intensive, and intensive systems of rainfed cropping (from Ruthenberg, 1971). Variations in local environmental and socioeconomic conditions make it difficult to discern any general relation between input per hectare and cultivation frequency. An increase in frequency should reduce input per hectare since with intensification the labor requirement for precultivation tasks—clearing and burning—will decline. Against this, the input into land preparation and weed control will tend to rise with intensity of cultivation: in shifting cultivation systems land preparation, in the sense of significant soil disturbance with tools, may be almost zero, and crop areas are often abandoned as weed infestation becomes serious. Even the contrast in input per hectare between hoe and draft animal cultivation is not clear. It should be kept in mind that the use of ox-drawn implements other than the plough and harrow is minimal in tropical crop production; hence it is normally only in the land preparation phase that draft animals can make any appreciable saving in

TABLE 5.5
Estimates of labor input per hectare for shifting cultivation, semi-intensive and intensive systems of rainfed cropping (excluding transport and processing)

Region	Crop	Hoe/draft animal	Labor input (hr ha^{-1})
a. *Shifting cultivation*			
Zaire	rice/cassava/ bananas	hoe	884
Kalimantan*	upland rice	hoe	620–880
Guatemala**			
(mature forest)	corn, etc.	hoe	689
(forest 4–7 yr)	corn, etc.	hoe	491
(forest 1–3 yr)	corn, etc.	hoe	447
b. *Semi-intensive*			
Tanzania	rice	hoe	1,250
	cotton/corn	hoe	842
	cotton/sorghum	hoe	614
Senegal	peanuts	hoe	755
Upper Volta	millet	hoe	476
	corn	hoe	468
Senegal	peanuts	draft	456
Tanzania	cotton	draft	1,520
c. *Intensive*			
Senegal	millet	hoe	356
Cameroon	soybean	hoe	552
Madagascar	cassava	hoe	446
Senegal	millet	draft	224
Taiwan	sweet potato	draft	804
Thailand	rainfed wet rice	draft	656

SOURCES: *From Freeman (1955); **from Carter (1969); remainder from Ruthenberg (1971).

human labor input per hectare. Table 5.5 suggests that the average level of input for rainfed cropping systems is in the range of 500 to 1,000 hr ha^{-1}.

The association of irrigation with a higher input per hectare is seen clearly in the figures given in Table 5.6 from Boserup (1965), which contrast rainfed and irrigated production of wheat, sorghum, and gram at Indian locations.

TABLE 5.6
Labor input for rainfed and irrigated cropping

Location	Crop	Input (man-days ha^{-1})	
		Rainfed	Irrigated
Bombay	Wheat	39	106
Punjab	Wheat	33	65
Bombay	Sorghum	25	77
Madras	Sorghum	39	111
Bombay	Gram	33	55

For the crops and locations given in Table 5.6, the average increase in labor input per hectare with irrigation was 145 percent.

■ 5.6.5 *Input per unit of crop production.* In comparison with mechanized cropping the productivity of human labor, unsupported by an array of machines, is naturally very low. Within nonmechanized tropical systems, an increase in labor input per unit of crop produced or, to use a more conventional form of expression, a decline in crop production per unit labor input, would be expected with increasing frequency of cultivation, since one of the essential principles of this shift is progressive substitution of labor for land. Boserup (1965) comes to the conclusion that "output per manhour is more likely to decline than to increase" in the circumstances.

Within any one cropping system, there will of course be a reduction in crop output per unit labor input with increasing input per hectare. Haswell (1973) gives examples of the marginal productivity of labor for corn, upland rice and wet rice in Mindanao, South Philippines. The capacity of wet rice systems to continue to respond in output to increasing labor inputs, which is shown by Haswell's data, has already been mentioned (Geertz, 1963).

■ 5.7 Labor potential of the farm family

■ 5.7.1 *Division of labor.* In section 5.6 labor input was expressed in simple terms as hours of adult male equivalent labor devoted to crop production. On the tropical family farm, all but the very old, the very young, and the incapacitated are likely to contribute to the total farm effort, not merely in crop production tasks but in other activities such as care of stock and food processing and preparation. Females are much concerned with the latter, but in some instances they do as much field work as the males, or more.

In traditional agricultural societies, custom often determines the division of labor within the family. Ruthenberg (1971) distinguishes three modes of labor division between the sexes in shifting cultivation and semi-intensive farming systems:

a. Division of tasks. For example, it is customary for men only to undertake the heavy tasks of vegetation clearing and burning.
b. Division of cropfields. After the male task of clearing and preparation is complete, the individual members of the family may be responsible for crop production tasks on specific areas.
c. Division of crops. In mixed subsistence–cash crop systems, males are more likely to be responsible for the cash crop and females for the food crops.

■ 5.7.2 *Hired labor.* While permanent hired labor may form a significant component of the total farm work force in intensive cash crop production systems (e.g., irrigated vegetable-growing) and in countries with a high proportion of landless laborers (e.g., India), on the family subsistence farm labor is normally hired only for periods of peak demand. Norman (1970) notes in West Africa that hired labor may be retained during off-peak periods merely so that the head of the household can enjoy leisure, and be seen by his peers to enjoy it, thus gaining social prestige.

■ 5.7.3 *Incapacity.* Two characteristic limitations to the productivity of labor in tropical agriculture have already been noted briefly: health and seasonal food supply. In developing countries of low food output and poor health services, the capacity of the work force can be seriously reduced by ill-health, which may take the form of chronic debilitation (through, for instance, protein-calorie deficiency, vitamin deficiency, hookworm, etc.) or of intermittent incapacitation through poorly controlled disease epidemics (FAO, 1966).

In addition, in climatic regimes which impose a single cropping season and a concentrated crop harvest period, seasonal variation in food supply may influence work output (Johnston, 1958). This would be serious even if work demand were relatively constant throughout the year, but its effect is aggravated by the fact that the second half of the post-harvest year, the food scarcity period, is when field work for the next season's crop is in progress.

REFERENCES

Allan, W. 1965. *The African husbandman.* Edinburgh and London: Oliver and Boyd.

Boserup, E. 1965. *The conditions of agricultural growth.* Chicago: Aldine.

Charreau, C. 1974. *Systems of cropping in the dry tropical zone of West Africa with special reference to Senegal.* Hyderabad, India: International Workshop on Farming Systems. ICRISAT.

Carter, W. E. 1969. *New lands and old traditions. Kekchi cultivators in the Guatemalan lowlands.* Center for Latin American Studies. Gainesville, Florida: University of Florida Press.

Clark, C., and Haswell, M. 1970. *The economics of subsistence agriculture.* London: Macmillan.

Food and Agriculture Organization. 1966. *Nutrition and working efficiency.* Freedom from Hunger Campaign Basic Studies no. 5.

Freeman, J. D. 1955. *Iban agriculture.* Colonial Research Studies. London: H. M. Stationery Office.

Geertz, C. 1963. *Agricultural involution. The process of ecological change in Indonesia.* Los Angeles: University of California Press.

Grigg, D. B. 1974. *The agricultural systems of the world. An evolutionary approach.* London: Cambridge University Press.

Haswell, M. R. 1973. *Tropical farming economics.* London: Longman.

Johnston, B. F. 1958. *The staple food economies of western tropical Africa.* Stanford, California: Stanford University Press.

Kampen, J. 1974. *Soil and water conservation and management in farming systems research for the semi-arid tropics.* Hyderabad, India: International Workshop in Farming Systems. ICRISAT.

Norman, D. W. 1968. How hard do Nigerian farmers work? *Samaru Agricultural Newsletter* 9:28.

Norman, D. W. 1970. Labor inputs of farmers: a case study of the Zaria province of the North-Central State of Nigeria. *Samaru Research Bulletin* 116:3.

Penny, D. H. 1969. Growth of "economic-mindedness" among small farmers in North Sumatra, Indonesia. In: Wharton, C. R., ed. *Subsistence agriculture and economic development.* Chicago: Aldine.

Ruthenberg, H. 1971. *Farming systems in the tropics.* Oxford: Clarendon Press.

Ryan, J. G. 1974. *Socio-economic aspects of agricultural development in the semi-arid tropics.* Hyderabad, India: International Workshop on Farming Systems. ICRISAT.

Wharton, C. R. 1969. Subsistence agriculture, concepts and scope. In: Wharton, C. R., ed. *Subsistence agriculture and economic development.* Chicago: Aldine.

Chapter 6

Shifting Cultivation Systems: General Aspects

■ 6.1 Definitions and Nomenclature

■ 6.1.1 The definition of shifting cultivation is imprecise. In its broadest connotation, the term may be applied to all systems of rainfed cropping with annuals, biennials, or short-lived perennials in which a cropping period alternates with a longer rest or fallow period, during which the abandoned crop area is recolonized by native herbaceous, shrub, or tree species or by adventive species that find the ecological conditions favorable.

There is a continuous range of cropping systems of different cultivation frequency, and the boundary between shifting cultivation and semi-intensive or "semipermanent" rainfed cropping (Ruthenberg's term, 1971) is arbitrary. Sanchez (1973) uses the term "shifting cultivation" to denote cropping systems in which the fallow period exceeds the crop period; that is, when cultivation frequency is less than 50 percent. Ruthenberg (1971) nominates a cultivation frequency value of 30 percent to demarcate shifting cultivation from "semipermanent" cropping. In Allan's classification (1965), which classifies land not by farming system but by its ability to support stable cropping systems of a given cultivation frequency, the term shifting cultivation is reserved for systems with a cultivation frequency of 10 percent or less, and those of 10 to 33 percent he designates "recurrent cultivation." Allan actually expresses his class values in terms of a "land-use factor" which is the reciprocal of cultivation frequency; i.e., cropping period plus fallow period divided by cropping period. In the present text, Ruthenberg's limit of a cultivation frequency of not more than 30 percent is adopted as the dividing line between shifting cultivation and semi-intensive rainfed cropping.

In practice, it is difficult to determine average cultivation frequency for a shifting cultivation system operating in a given area, for a number of reasons. First, it is unusual for shifting cultivators to adhere to a strict time sequence of crop and fallow. Second, at low cultivation frequencies the area of potentially arable land that may form part of the fallow sequence cannot be accurately determined. Third, frequency of cultivation will vary between small areas in accordance with inequalities in the environment: soil, slope, or aspect, for example. Finally, cultivation frequency on specified areas is also likely to vary within a single farm unit, generally in accordance with distance from the domicile.

A wide assortment of names, some internationally recognized, some national, and some purely local, are applied to shifting cultivation in general. Two terms in common international use are "slash-and-burn" and "swidden" cultivation; "swidden," from an Old English word meaning a clearing in the forest, is popular with anthropologists. Well-known regional terms include *milpa* and *conuco* in Central America and *kaingin* in the Philippines. Conklin (1957) and Spencer (1966) treat of nomenclature in detail.

■ 6.1.2 There have been a number of attempts to classify subtypes of shifting cultivation on the basis either of cultivation frequency or character of the fallow vegetation, criteria that are to a substantial degree linked. Mention has already been made of Allan's distinction between shifting and recurrent cultivation. Morgan (1969) distinguishes between shifting cultivation systems, in which cropfield boundaries are not permanent from one crop phase to the next, and "rotational bush fallow" systems, which are associated with a greater cultivation frequency and where cropfield boundaries remain more or less permanent. Boserup (1965) separates "forest fallow" systems with rest periods of twenty years or more, during which time a secondary forest community can reestablish itself, from "bush fallow" systems with rest periods of 6 to 10 years, which may be only long enough for the regeneration of herbaceous and scrub communities. The use of fallow-phase vegetation characteristics to categorize shifting cultivation types is taken further by Spencer (1966). Within the definition of shifting cultivation adopted here, no distinctions will be made between subtypes.

■ 6.1.3 As has already been noted, the numerical concept of cultivation frequency, or its reciprocal, the land-use factor, does not imply a regular crop–fallow time sequence, though normally as cultivation frequency increases there is an approach toward regularity. Nor does it necessarily represent only the estimated mean of a population of simple crop–fallow sequences. In Africa, for example, double-hierarchy sequences are found: for instance, 2 crop – 6 fallow – 6 crop – 12 fallow, or 3 crop – 5 fallow – 3 crop – 5 fallow – 3 crop – 20 fallow, the figures signifying years (Allan, 1965). The long fallow period in the latter sequence may involve a domicile shift by a whole group of farmers from a depleted cropping area to new land some distance away. Furthermore, of course, different sequences may have the same cultivation frequency: e.g., 2 crop – 10 fallow and 1 crop – 5 fallow. Table 6.1 gives data on shifting cultivation patterns in five African nations, obtained from a continent-wide FAO questionnaire. A further summary is given by Nye and Greenland (1960).

■ 6.2 Historical and geographical

■ 6.2.1 Shifting cultivation, though now largely confined to the less-developed areas of the tropics, has a long historical background in both temperate and tropical zones. It is most likely that it was the first system of crop agriculture; certainly there was no alternative method of maintaining soil fertility known to the earliest cultivators. Throughout the history of agricultural settlement it has been the characteristic system of peoples colonizing new areas in situations where labor rather than land was the limiting production factor and only hand tools were available. Crop agriculture as it spread from Southwest Asia into Europe via the Danube basin was by shifting cultivation methods, and in the northern European forests the system survived into the nineteenth century. Even the settlement of the southern U.S.A. often involved a type of shifting cultivation: forest land was cleared by ring barking and felling, cropped to corn or cotton, and abandoned to revert to scrub.

As the rural population of settled areas increased, forest and grassland were permanently cleared and a predominantly shifting agriculture gave place in time to systems of greater cultivation frequency on the better land areas. Shifting cultivation became a residual farming system practiced on less productive land, frequently by groups socially or ethnically distinct from the more "progressive" societies.

The history of the spread of shifting cultivation is outlined by Grigg (1974), and a more detailed account for Southeast Asia is provided by Spencer (1966). The current pattern for tropical Africa, Asia, and America is described by Morgan (1969), Spencer (1966), and Watters (1971), respectively. It was estimated by FAO that in 1957 some 200 million people in the world were dependent on shifting cultivation systems for sustenance; they occupied an estimated total area, cropland and fallow, of 33 million km^2.

■ 6.2.2 Within the tropics, shifting cultivation systems are of greatest importance in Africa, though it is hard to arrive at any accurate figure of the number of people who practice it, or of the area over which it is practiced, because the range of cultivation frequency is so wide and continuous. Estimates depend merely on the designation of what constitutes shifting cultivation and what semi-intensive rainfed cropping.

Tropical Africa has a long history of agricultural settlement, but over much of it population density has never built up to the levels found, for example, in South and Southeast Asia. Furthermore, crop area per farming unit was limited by the fact that draft cattle and the ox-drawn plough only became common in quite recent times. Morgan (1969) has classified and mapped the main types of shifting cultivation and "rotational bush-fallowing," to use his own terminology: the former predominates in east, central, and south tropical Africa and the latter in West Africa. The general picture in Africa, however, is that of a great range of shifting and semi-intensive hoe cultivation systems diffused over the whole of the tropic zone, with some areas of intensive development.

■ 6.2.3 Spencer (1966) has estimated that in South and Southeast Asia about 50 million people are shifting cultivators, cropping each year some 10 to 18 million ha within a total utilization area of 100 million ha. However, this represents only about 7 percent of the rural population, and the total area of crop and fallow land utilized is less than half that of the land under permanent cultivation. It seems probable that in southeast Asia at least the original cropping systems were based on noncereal carbohydrate crops—taro, yams, plantains—which were gradually displaced by seed crops diffusing from India and China, that is, upland grains such as millets, upland rice, and pulses. Wet rice cultivation appears to be a later development, and with the gradual development of rice cropping in lowland areas shifting cultivation retreated to

TABLE 6.1
Some types of shifting cultivation in Africa

Country	Area under shifting cultivation (ha)	Rural population per ha of cropland	Average crop yields per ha		Typical cropping systems		
			Crop	Yield	Crops	Crop phase (yrs)	Fallow phase (yrs)
Congo Peoples Republic	90,000	0.9	Plantains	8–15 t	Pumpkins, corn, yams, peanuts, cassava	4	6–10
			Cassava	8–10 t			
			Sweet potatoes	5 t			
			Peanuts	6–800 kg	Peanuts	1–2	4– 6
			Corn	5–600 kg			
					Tobacco, pumpkins, sweet potato, yams, cassava	1–3	8–10
					Pumpkins, cassava peanuts	2 1	2– 3 then 8–10
Dahomey	900,000	1.5	Corn	300–1,000 kg	Corn, peanuts, yams, cotton	2–3	3–18
			Peanuts	4–800 kg			
			Yams	6–10 t			
			Cassava	5–9 t	Yams, beans, corn, cotton, sorghum	2–3	3–10
Liberia	75,000	0.2	Rice	600 kg	Cassava, corn, rice, yams, sweet potato	1½	3–5
			Peanuts	600 kg			
			Cassava	5 t			

					Rice, corn, cassava, peanuts, sweet potato	1½	3–7
					Rice, corn, cassava, okra	1½	2– 7
Malawi	?	1.3	Corn Millet Cassava	500 kg 2–300 kg 5–30 t	Corn, millet, sorghum, cassava	1–3	6–10
					Corn, millet, sorghum, cassava	1–3	20–25
Zambia	1,900,000	0.7		No reply	Cassava, sweet potatoes, corn	2–3	8
					Millet, beans, peanuts	2	10
					Corn	5	15

SOURCE: Braun (1974).

sites where topography and soils were unsuitable for paddy (Grigg, 1974). On the Indian subcontinent, the general pressure of population increase in areas too dry for extensive wet rice cropping has led to the displacement of shifting cultivation by intensive rainfed cropping systems. The current situation in Asia therefore differs markedly from that in Africa: instead of a diffuse range of cropping systems of varying cultivation frequency the characteristic pattern is a fairly sharp contrast between intensive cultivation (with draft animals) in the lowlands and shifting cultivation on hill land.

■ 6.2.4 In tropical America shifting cultivation was practiced at least before 1,000 B.C. It was based on corn, beans, and squash in the drier tropical areas of Mexico and on tubers, cassava and sweet potatoes particularly, in the wetter equatorial lowlands further south. Throughout the region, shifting cultivation is associated more with Indian ethnic groups than with peoples of European origin, and with the less favorable and more remote localities. However, Watters (1971) notes that frequently the traditional systems developed by Indians are more stable than those of European peoples, who are still in an "experimental" stage of crop system development. Although shifting cultivation (as defined by cultivation frequency of less than 30 percent) is in tropical America characteristically a lowland cropping system, utilizing the still extensive tropical forests, it is found over a wide range of altitudes and rainfall regimes. However, in high-altitude regions the crop–fallow rotation is normally shorter and many of the cropping systems would be classified as semi-intensive.

■ 6.2.5 Before the linking of the Old World and the New in the fifteenth century, the crops grown in shifting cultivation systems were either indigenous to the locality or had spread slowly into it by cultural diffusion. One of the more remarkable features of the subsequent history of shifting cultivation has been the rapid assimilation of New World crops into Old World cropping systems. Particularly important has been the adoption in Africa and Asia of corn, cassava, and sweet potatoes, three energy staples readily adaptable to the ecological conditions of shifting cultivation and appropriate to its largely subsistence character. Perhaps the most striking example is the great dependence of the highland peoples of New Guinea on the alien sweet potato, which, owing to its broad temperature adaptation, has permitted shifting cultivation over a wide altitudinal range (Brookfield and Brown, 1963).

■ 6.3 Characteristic general features of shifting cultivation

■ 6.3.1 Shifting cultivation may or may not involve a shift of domicile. Attempts have been made to confine the term to systems in which the settlement itself moves at intervals, but the distinction has never become established. However, it is broadly true to say that shifting cultivators are less sedentary than farmers operating more intensive systems. There are two distinguishable poles of domicile shift, one limited in time and space, the other taking place over longer distances and intervals. The word poles is used advisedly since a range of intermediate situations is found.

If under the particular cultivation frequency adopted the farmer finds it difficult to sustain fertility levels and hence crop yield within an area close to home, new areas further out will be cleared and cropped. This will involve more walking and transport of crops to storage sites, and in time it may prove energetically less demanding to shift the domicile over a relatively short distance to the newer cropping area. On the other hand, a village or settlement group, after a fairly long period of utilization of a given region for cropping, may find that the productivity of the whole commandable area has fallen to intolerable levels. This may lead to a wholesale movement of the group or village to a new site some distance away. As indicated in section 6.1.3., a major shift of this nature may coincide with a long fallow phase in complex crop–fallow time sequences.

■ 6.3.2 Shifting cultivation is characteristically undertaken with hand tools and draft animal power is very uncommon, for a number of reasons. First, the cultivators may be too poor to buy cattle; second, in wet forest regions maintenance of the health and diet of cattle may be difficult; third, the degree of land clearance achieved on cropfields is usually inadequate for ploughing except in grass savannah environments; and fourth, cropping is often undertaken on slopes too steep for draft animals to work on.

We can distinguish two main types of shifting cultivation on the basis of the hand tools used: digging stick systems and hoe systems. In forest, where the ground vegetation layer is negligible or at least not dominated by grasses, the surface soil is often scarcely disturbed after burning, and seeds or tubers are planted directly in the ash with a digging stick. On the other hand, in savannah grassland hoes are necessary to break up the surface soil and to chop and dispose of the sod. Furthermore, in systems where the cropping break is extended and weed control is a major input, hoes are needed; whereas in systems where weed problems are largely evaded by abandoning the cropfield, only a cutlass or machete may be required to control light infestations.

■ 6.3.3 Shifting cultivation is overwhelmingly a subsistence form of agriculture. Over very large areas of tropical America and Asia it is the cropping system practiced by the disadvantaged sectors of the rural population in the more remote localities, and the absence of a road network in dense forest or hilly terrain precludes the development of markets for cash crops. It is perhaps significant that one of the most highly developed cash crop shifting cultivation systems is opium growing (in conjunction with corn and upland rice) in the hills of North Thailand, Burma, and Laos. Opium is a crop of very high value per unit weight and volume, and marketing is physically simple if legally hazardous.

■ 6.3.4 In African and Asian shifting cultivation systems at least, farm livestock play a minor role. Nonruminant animals—chickens and pigs, fed largely on carbohydrate grains and tubers and making up the remainder of their dietary requirement by free-range feeding—are more common than ruminants, though in savannah areas cattle may be accumulated largely as a store of wealth. In the savannah areas of tropical America, however, subsistence food production by shifting cultivation is often associated with the ownership of substantial herds of cattle.

■ 6.3.5 The simple tools used, the apparent crudity of a technique in which crops are grown in roughly cleared temporary sites, the lack of integration of crops and livestock, and the normally subordinate position of the farming population in the rural society of which they form a part—all these characteristics tend to create the impression that shifting cultivation is a "primitive" form of agriculture. It is primitive in the sense of being primal, a direct evolution from the earliest systems of crop agriculture, and also in the sense of having benefited least from modern technological advances, but in its execution it often demonstrates a subtle technical adaptation to the environment. Later in this chapter, when we consider the organization of cropping in time and space, some of the more complex manifestations of shifting cultivation technology will be discussed.

■ 6.4 Forest and savannah

■ 6.4.1 As a basis for understanding relations between cropping patterns and fallow-phase vegetation in shifting cultivation systems, a brief treatment of the ecological balance of savannah and forest is necessary. Figure 4.1 illustrates the environmental and anthropogenic factors influencing this balance. It does not take into account true edaphic savannahs, in which the dominance of grass or sedge communities is normally associated with poorly aerated soils liable to flooding.

In general, dependent on local environmental factors but independent of man's influence, there is a shift from closed forest toward savannah with decreasing annual rainfall, increasing length of the dry period, and decreasing soil depth, ground water level, and water-holding capacity.

Furthermore, in any given environment, with increasing frequency of clearing, burning, and cultivation, there is a direct effect on vegetation succession which favors shorter and more shallow-rooted trees and scrub—secondary forest—at the expense of the climax primary forest, and favors grassland savannah at the expense of secondary forest. Initially, these changes occur as a result of differences between the vegetation complexes in respect of their ability to recover from being cut down and burned and their ability quickly to recolonize abandoned cropland. In time, however, intermittent cropping and the vegetation changes in themselves may lead to changes in the biogeochemical and hydrological characteristics of the soil and its surface, a deterioration that reinforces the successional trend.

■ 6.4.2 There are, naturally, numerous variations on and exceptions to the type pattern of succession enunciated above. Some subsidiary aspects of importance are mentioned below:

a. *Spread of fire.* When closed forest is cleared and burned, the fire extends little beyond the cleared area, for the standing forest is not combustible. With increasing frequency of clearing and burning, the forest is opened up, allowing light-demanding grasses to colonize the ground layers. When this occurs, the likelihood of fires extending beyond the cleared area increases, accelerating the trend toward grass dominance.

b. *Time of burning.* In undisturbed communities of scrub and grass, i.e., those not cut down for land clearing, burning in late dry season favors grass at the expense of scrub, whereas burning in early dry season does not. In late dry season, the grasses are either dead, if annuals, or dormant, if perennials, and are little affected by fire; the scrub on the other hand is beginning to grow on stored subsoil water in response to rising temperature and is vulnerable.

c. *Grazing.* Up to a point, increasing grazing pressure will favor the succession from forest to savannah, since young tree seedlings or suckers are more susceptible to defoliation and trampling than are grasses. Ruminant herds are a significant factor in maintaining savannah communities. On the other hand, in some semi-arid regions further heavy grazing pressure may favor the development of communities dominated by unpalatable or physically repellent scrub (e.g., spiny species). The process is to some degree autogenic: with more scrub and less grass, late dry season burning, which tends to reverse the process, becomes progressively more difficult.

d. *Erosion.* At any given cultivation frequency, the successional changes from forest to grassland are often accelerated on soil types that are erodible, or on slopes that encourage erosion, because the biogeochemical and hydrological degradation characteristic of the succession is augmented by the loss of topsoil.

■ 6.5 Clearing, burning, and seedbed preparation techniques

■ 6.5.1 The sequences of operation in shifting cultivation systems from standing fallow vegetation to crop seedbed vary greatly in accordance with the ecological environment and with the cultural pattern of the cultivator group. Miracle (1967) discerned twelve main types in the Congo basin alone, and these do not represent the full range across the tropical world. The more important of these sequences are described briefly below:

a. *Clear, burn, and plant* (forest). The classical system found throughout the world in closed forest, where the surface soil is friable and the ground flora sparse. No hoeing is needed and crops are planted directly in the ash.

b. *Clear, burn, hoe, and plant* (forest). A more laborious version of (a) above, where soil conditions dictate a cultivated seedbed made with hoes. This operation may precede the formation of mounds (see section 6.5.2) as a microenvironment for specific crops.

c. *Clear, plant, and burn* (forest). In some African situations crops—normally robust nonseed species such as plantain or cassava—are planted in the roughly cutover vegetation. After they are established, clearing operations are completed and the debris burned, with little danger to the crop. Labor inputs are thus spread over a longer period and ash nutrients are not leached out before crop roots can intercept them.

d. *Clear, allow to rot, and plant* (forest). In very high rainfall areas of tropical America, where burning is difficult, the cleared vegetation may be merely allowed to rot on the ground before planting (Sanchez, 1973). An interesting variant is found in areas of lower rainfall in the Sudan, where termites accomplish the disintegration of cut timber in about four months (Miracle, 1967).

e. *Double cycle clear-burn-plant* (forest). In this African system, the crop area is roughly cleared and burned for first-year crops and the operation is repeated more thoroughly in the second year. Bananas or plantains are planted on the first year—crops suitable for semicleared situations—and seed crops in the second year. Again, labor inputs are spread.

f. *Citamene systems* (forest). This term is applied to a group of African techniques, the distinctive character of which is that additional wood is cut from the surrounding forest and is brought on to the crop area to be burned, thereby increasing the ash nutrient supply. The area from which wood is cut may be 5 to 20 times as large as the crop area. The land outside the crop area is not fully cleared: small trees are cut at breast height and larger trees are lopped. Allan (1965) describes such systems in detail and distinguishes between "large-circle" and "small-circle" citamene systems according to whether the wood is stacked and burned in a single large heap or in a series of smaller heaps.

g. *Cut, burn, hoe, and plant* (savannah). The classical savannah clearance technique: in contrast to closed forest clearing, thorough hoeing is necessary to kill the grasses and to create a loose seedbed. This forms the basis for a number of different subsequent procedures of mounding, ridging, etc., before the crops are planted (see section 6.5.2).

h. *Burn, hoe, and plant* (savannah). In some savannah areas the standing dry grass is burned without prior cutting and the crop area is then hoed. The system is sometimes associated with mounded seedbeds. The savannah is normally burned in late dry season, but one variant is to delay burning until early wet season, by which means the new grass shoots are killed and reestablishment of grass in the cropfield diminished.

i. *Cut, hoe and burn, plant* (savannah). Under this system the cleared grass, cut shoots, and hoed roots, are collected into heaps which are covered with soil and allowed to rot or are burned and then covered with soil. These heaps form the basis of mounds.

■ 6.5.2 *Seedbed geometry.* Reference has been made above to the creation of mounds, beds, and ridges before planting. Although in many shifting cultivation systems, particularly those associated with closed forest, crops are planted on the flat, there are numerous instances, both in savannah and forest but more commonly in the former, where the surface soil is hoed into mounds, ridges, or beds before planting. Such techniques are more characteristic of Africa and Oceania than elsewhere. Another method is found in New Guinea, where drainage trenches are dug in a crisscross fashion and the excavated soil mounded on the beds thus created. The range of surface soil geometries and their function is discussed by Denevan and Turner (1974). The advantages of creating raised microenvironments in which to plant specific crops include:

a. *Soil fertility.* Topsoil, richer in nutrients and organic matter, is concentrated. This obviously involves a reduction in the nutrient content of the surrounding areas from whence the topsoil is hoed, but in low-fertility soils the threshold reached in the mounds may permit specific fertility-demanding crops to be grown. Nutrient and organic matter content of the mound may be augmented by collecting vegetation, burning it or allowing it to rot, and forming a mound over it. In savannah environments, improved aeration in the mound hastens the mineralization of organic nitrogen from carbonaceous organic matter.

b. *Soil temperature.* At high altitudes the microtopographic pattern of mounds may help crops planted on them to avoid low-temperature checks or damage. Such temperature differences have been noted in the highlands of New Guinea, where sweet potatoes are grown.

c. *Soil water.* In high rainfall or poor drainage conditions, mounds, ridges, or beds are well-drained and if they are of high organic matter content (particularly with rotted grass at the base of the mound), soil water retention is improved. This not only assists crop growth, but may also be important for the keeping quality of tuber crops that are allowed to remain in the ground after maturation and are harvested as required.

d. *Soil texture.* The loose texture of the mound encourages root proliferation in "tight" soils, helps the development of underground tubers, and makes their harvest easier. Individual tubers may be readily extracted from the mound as required without disturbing the remainder of the plant.

■ 6.6 Cropping tactics

■ 6.6.1 It is frequently stated that cropping patterns in shifting cultivation are characterized by the complex and varied assembly of crop species that are grown in a single clearing, and the classic example of the Hanunoo in the Philippines rainforest (Conklin, 1957) is almost universally quoted. This generalization is far too sweeping. It is true that under shifting cultivation in forest regions with favorable moisture conditions for most of the year crop diversity within a single clearing reaches its apogee. On the other hand, with increasing environmental limitation, toward a shorter wet season, lower fertility, or lower temperature determined by altitude, this diversity is progressively reduced. Monoculture under shifting cultivation is common in such environments: for example, corn in Mexico, upland rice in Southeast Asia, or sweet potatoes in the New Guinea highlands.

Perhaps a more generally applicable characteristic of shifting cultivation systems is that the cultivators attempt to exploit to the full the range of microenvironments within the cropfield. These may be of natural occurrence—termite mounds, old tree stumps, the shady and sunny sides of the clearing—or may be created, e.g., mounds, beds, and ridges. As Ruthenberg (1971) aptly puts it, "this agriculture apes the generalised diversity of the jungle that it temporarily replaces." To the casual observer, a cropfield in rainforest gives the impression of unplanned confusion and invites the epithet "primitive." Nothing could be further from the truth: the cultivator is using his intimate knowledge of the environment and of the requirements of his several crops to maximize production per unit area, time, or labor input.

■ 6.6.2 We may codify the principles of shifting cultivation cropping tactics in three dimensions: horizontal organization in space, vertical organization in space, and organization in time. In respect of temporal organization, we shall confine ourselves here to a single crop year; the patterns of year-to-year crop sequence under shifting cultivation will be examined in the next chapter.

a. *Horizontal organization in space.* Within the cropfield area, advantage is taken of environmental variation in the horizontal dimension. Shade-tolerant crops may be planted on the shady edge of the clearing. If it has dry and damp zones, for example at the top and bottom of a slope site, adapted crop species are planted accordingly. Fertility "islands," created by termite mounds or the localized ash concentration derived from a big tree stump, are planted to demanding crops.

b. *Vertical organization in space.* Crops of varying height, canopy structure, and radiation requirements may be planted together to exploit fully the light profile, and possibly also through different rooting habits the soil water and nutrient profile. Microenvironments differing in soil water, temperature, texture and fertility level in a vertical dimension are created in mounds, ridges, and beds: one crop type may be planted on top of the mound, others on the sides and yet others on the flat areas between. Climbing crops needing support, e.g., yams, beans, are planted around tree trunks or at the base of rigid upright crops.

c. *Organization in time.* The time sequence of crop planting may be complex and extended, to exploit fully seasonal variations in temperature and soil water supply and to maximize the use of solar radiation. The more or less continuously maintained crop canopy also protects the soil surface from erosion and excessively high temperatures.

■ 6.6.3 In the above summary of potential cropping tactics, that is, tactics which may be employed if the environment permits, their rationale is expressed only in terms of adaptation to variation in physical and biological factors. There are other advantages to complex cropping:

a. The risk of pests and diseases reducing total food supply is said to be minimized, though the evidence for this is scanty.
b. The risk of intermittent adverse weather, e.g., drought spells, reducing total food supply is minimized.
c. A diverse and nutritionally adequate diet is assured.
d. Extended crop harvest reduces the necessity for storage, often hazardous in climates with a prolonged rainy period.
e. The labor input of planting and harvesting is spread.

■ 6.6.4 Hitherto in this section we have been examining the potential for complexity within a single clearing. The zonation of cropping pattern in accordance with distance from the house may also become important, though these contrasts become more apparent in semi-intensive and intensive cropping. Cropped areas around the house, normally more intensively cultivated than the outlying clearings, are devoted to vegetables and fruit and to minor nonfood crops such as tobacco, cotton for fiber and gourds for culinary vessels. Where livestock are kept, the fertility of the home garden is often sustained by animal manure.

REFERENCES

Allan, W. 1965. *The African husbandman.* Edinburgh and London: Oliver and Boyd.

Boserup, E. 1965. *The conditions of agricultural growth.* Chicago: Aldine.

Braun, H. 1974. Shifting cultivation in Africa (evaluation of questionnaires). In: *Shifting cultivation and soil conservation in Africa.* FAO Soils Bulletin no. 24. Rome.

Brookfield, H. C., and Brown, P. 1963. *Struggle for land.* Melbourne: Oxford University Press.

Conklin, H. A. 1957. *Hanunoo agriculture. A report on an integral system of shifting cultivation in the Philippines.* FAO Forestry Development Paper no. 12. Rome.

Denevan, W. M., and Turner, B. L. 1974. Forms, function and associations of raised fields in the old world tropics. *Journal of Tropical Geography* 39:24.

Grigg, D. B. 1974. *The agricultural systems of the world. An evolutionary approach.* London: Cambridge University Press.

Miracle, M. P. 1967. *Agriculture in the Congo basin.* Madison: University of Wisconsin Press.

Morgan, W. B. 1969. Peasant agriculture in tropical Africa. In: Thomas, M. F., and Whittington, G. W., eds. *Environment and land use in Africa.* London: Methuen.

Nye, P. H., and Greenland, D. J. 1960. *The soil under shifting cultivation.* Harpenden, England: Commonwealth Bureau of Soils. Technical Communication no. 51.

Ruthenberg, H. 1971. *Farming systems in the tropics.* Oxford: Clarendon Press.

Sanchez, P. 1973. Soil management under shifting cultivation. In: Sanchez, P., ed. *A review of soils research in tropical Latin America.* North Carolina Agricultural Experiment Station Technical Bulletin no. 219.

Spencer, J. A. 1966. *Shifting cultivation in south-eastern Asia.* Berkeley and Los Angeles: University of California Press.

Watters, R. F. 1971. *Shifting cultivation in Latin America.* FAO Forestry Development Paper no. 17. Rome.

Chapter 7

Shifting Cultivation Systems: Biogeochemical Aspects

■ 7.1 INTRODUCTION

■ 7.1.1 Much of the research into nutrient cycling in soil and vegetation under shifting cultivation has been carried out in Africa. The work of Nye and Greenland and their colleagues and predecessors has been most ably summarized in a review which, although seventeen years old at the time of writing this book, remains the standard text on the subject (Nye and Greenland, 1960). More recently, investigations in Latin America have increased in tempo; the results have been reviewed by Sanchez (1973). Recent work in Africa is referred to by Vine (1968) and Moss (1969), and in the FAO bulletin on shifting cultivation and soil conservation (see reference to Mouttapa, 1974). There are few published data from Asia.

■ 7.1.2 In the ensuing sections, the biogeochemistry of the individual phases of shifting cultivation cycles will be considered seriatim, beginning with nutrient cycling during the late phase of the fallow, when the vegetation is relatively stable, then proceeding to the effects of clearing and burning, the cropping phase, and finally the recovery of fallow vegetation after cropping has ceased. In conclusion, the more general long-term biogeochemical effects of a sequence of shifting cultivation cycles are assessed.

■ 7.2 THE FALLOW PHASE UNDER STEADY STATE CONDITIONS

■ 7.2.1 The title of this section is slightly inaccurate, since by definition a steady state is characteristic only of climax communities, and most of the data and principles presented below concern vegetation in a subclimax or plagioclimax situation following an interruption to succession by cultivation. However, the phrase steady state is here used, albeit loosely, to describe the late stages of a fallow period, when ecological and biogeochemical changes from year to year are less rapid than in the early phases of recovery after a cropping period. A simple model of the nutrient cycle in natural vegetation is given in Figure 7.1.

■ 7.2.2 *The vegetation store in forest.* Sanchez (1973) summarizes the range of published data from forests in the Congo, Ghana,

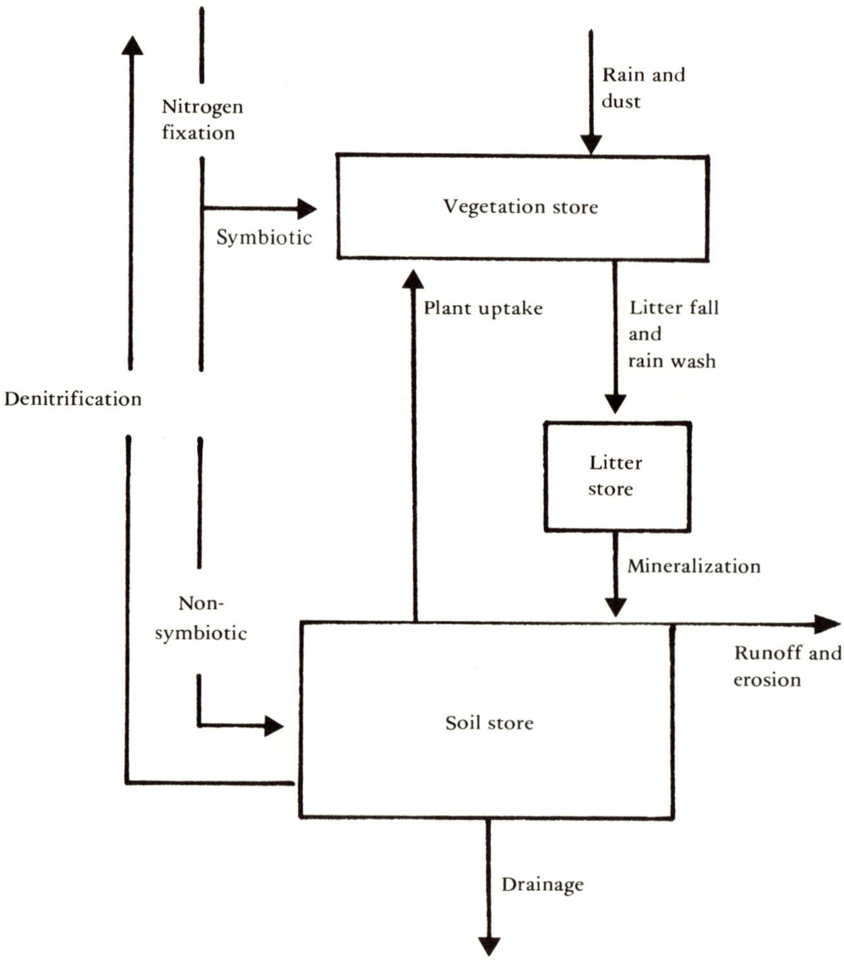

Figure 7.1. The nutrient cycle in forest (Nye and Greenland, 1960).

Panama, and Puerto Rico on dry matter and nutrient content in the biomass:

Dry matter	200– 400 t ha^{-1}
Nitrogen	700–2,050 kg ha^{-1}
Phosphorus	33– 137 kg ha^{-1}
Potassium	600–1,000 kg ha^{-1}
Calcium	650–2,750 kg ha^{-1}
Magnesium	400–3,900 kg ha^{-1}

A more detailed summary of data from secondary forests in Africa, partitioning the biomass between the morphological components of trees, is shown in Table 7.1. These figures also include litter.

TABLE 7.1
The vegetation store in forest

Location and rainfall (mm)	Years from last crop	Component	Dry matter (t ha^{-1})	Nutrients (kg ha^{-1}) N	P	K
Kade, Ghana (1,650)	40	leaves	25	480	32	82
		wood	308	1,316	91	615
		litter	2	35	1	10
		total tops	335	1,831	125	707
		roots	25	213	11	87
Yangambi, Congo (1,850)	18	leaves	6	145	7	81
		wood	133	338	64	316
		litter	6	75	2	8
		total tops	145	558	73	405
		roots	31	141	35	190
Kumasi, Ghana (1,500)	20	leaves	6	141	8	48
		wood	112	358	28	336
		litter	6	73	3	25
		total tops	124	572	39	409
		roots	?	?	?	?
Yangambi, Congo (1,850)	5	leaves	6	125	7	80
		wood	74	186	13	250
		litter	6	80	3	15
		total tops	87	391	23	345
		roots	26	179	8	113

SOURCE: Adapted from Nye and Greenland (1960).

Table 7.1 shows that the greatest proportion of dry matter and nutrients in the total vegetation store is in structural tissue of wood and bark. In these high-rainfall forests (1,500 to 1,800 mm) the litter store is quite small. The relation displayed between age of the community and dry matter and nutrient yields reflects adversely on the title to this section, and demonstrates a steady accumulation as the forest matures.

■ 7.2.3 *The vegetation store in savannah.* Savannah vegetation, dominated by grasses that do not form permanent structural tissue, accumulates much smaller amounts of dry matter and nutrients in the biomass. Table 7.2 shows the range of annual forage production in tropical savannahs. If the community is burned or grazed hard, the

TABLE 7.2
Annual production of forage in tropical savannahs

Location	Rainfall (mm)	Yield (t ha^{-1})	Vegetation
Ejura, Ghana	1,500	8.7	Derived savannah
Venezuela llanos	1,300	2.4–4.0	*Trachypogon* dominant*
Olokemeji, Nigeria	1,200	6.8	Derived savannah
Varanasi, India	1,100	7.4	Secondary grasslands
Shika, Nigeria	1,100	3.4	Guinea savannah
Kivu, Albert Park	900	3.2–6.3	*Themeda-Heteropogon*
Katherine, N. Australia	900	1.4	*Themeda**
Gir Forest, India	800	4.1	Shrub savannah, *Dichanthium*
Gir Forest, India	800	2.9	Tree savannah, *Themeda*
Serengeti, Tanzania	700	5.2	Tall grass savannah
Matopos, Rhodesia	650	1.4	*Themeda-Heteropogon*
Rimé, Tchad	300	1.2–1.6	*Brachiaria deflexa*

SOURCE: Adapted from Bourlière and Hadley (1970).
*Low fertility locations.

carry-over of the year's aboveground production is negligible. In a protected savannah woodland in North Australia, the author estimated the standing biomass of ungrazed and unburned grassland at about 150 percent of the average annual production, i.e., a carry-over of 50 percent. In the llanos of Venezuela, the standing biomass of protected grassland represents 2.5 to 4 times annual production.

Table 7.3 shows that in open woodland savannah in Ghana total aboveground biomass and nutrient store is dominated by the structural tissue of the trees. The last example in Table 7.3 is from a low-rainfall area (Rhodesia); in a similar rainfall zone in North Australia the average dry matter yield of forage is 1,500 to 2,000 kg ha^{-1}, nitrogen yield 8 kg ha^{-1}, and phosphorus yield 0.5 kg ha^{-1}.

■ 7.2.4 *The soil nutrient store in forest and savannah.* Table 7.4 gives an indication of nutrient store in the top 30 cm of soil under forest and savannah in Africa. The figures are quite approximate and a standard bulk density was assumed.

In the absence of information on rates of mineralization of nitrogen and phosphorus, these data are of limited value, but they illustrate the general magnitude of the soil store in relation to the vegetation store (Tables 7.1 and 7.3). Nye and Greenland state that total phosphorus in

TABLE 7.3
The vegetation store in savannah

Location and rainfall (mm)	Years from last crop	Component	Dry matter (t ha^{-1})	Nutrients N	P	K
Ejura, Ghana* (1,500)	20	grass tops	7.6	22	7	41
		grass roots	3.8	13	3	11
		tree leaves	0.5	7	1	5
		tree wood	53.8	93	13	140
		litter	1.2	4	1	4
Ejura, Ghana** (1,500)		leaves, litter	3.9	17	6	35
		roots, stolons	8.8	29	13	71
Marandellas, Rhodesia (900)		grass	1.4	9	1	18
		other plants	1.0	13	1	12

SOURCE: Adapted from Nye and Greenland (1960).
**Andropogon* spp. dominant, open woodland.
***Imperata cylindrica* dominant.

soils of Ghana, Liberia, and the Congo is 20 to 100 times available phosphorus (Truog). At any one time the concentration of nitrate-N will be low, but this may merely signify that the sinks for nitrate are stronger than the sources and that it is being taken up as fast as it is mineralized. Alternatively, the nitrification capacity of the soil under forest may be low.

Although the limited range of data in Table 7.4 does not show any clear relation between the magnitude of the soil nutrient store and the type of vegetation, it is generally accepted that total soil nutrients are lower under rainforest in wet climates than in semi-evergreen forest developed under lower rainfall. The comparison between semi-evergreen forest and savannah is even less clear, though the nutrient store in savannah is likely to be smaller.

■ 7.2.5 *Nutrient return from vegetation to soil in forest.* The general pattern of nutrient cycling in tropical forest is that of a limited total store of nutrients in the soil and vegetation being fairly rapidly cycled through the soil/plant system. The phases of the cycle are the return of nutrients to the soil surface from the vegetation, the decomposition of this material, and the subsequent nutrient uptake by the trees.

Table 7.5 shows that the annual rate of litter fall in tropical forests is much greater than that in subtropical or temperate forest. A par-

TABLE 7.4
The soil nutrient store (0–30 cm) under forest and savannah

			Nutrients (kg ha^{-1})				
Vegetation	Location	pH	N	P*	K**	Ca**	Mg**
Rainforest	Ghana	3.7–5.3	4,600	12	650	2,600	370
	Liberia	3.9–4.5	4,000	26	460	500	230
	Congo		2,200–3,300	19	360	100	50
Semi-evergreen forest	Ghana	6.2–6.4	6,700	15	250	4,700	500
	Ghana	4.8–5.8	8,100	21	430	3,000	?
	Nigeria	6.2–6.5	?	25	310	2,200	220
Dry forest and savannah	Ghana	6.9–7.2	7,200	36	460	6,400	970
	Zambia	4.9–5.3	2,700	?	710	600	20
	Ghana	6.4–7.1	1,800	22	190	2,900	380
	Ghana	6.4–6.9	3,700	4	320	2,800	5,700

SOURCE: Adapted from Nye and Greenland (1960).
*Available P, generally by Truog.
**Exchangeable.

TABLE 7.5
Annual rate of litter fall in forest

		DM	Nutrients (kg ha^{-1})				
Climate	Location	(t ha^{-1})	N	P	K	Ca	Mg
Tropical	Congo	14.6	130	4	104	114	43
Tropical	Congo	12.3	224	7	48	105	52
Tropical	Ghana	10.5	199	7	68	206	45
Subtropical	Australia	6.7	108	6	37	75	?
Temperate	N.E. USA	3.0	19	3	16	74	10

SOURCE: Nye and Greenland (1960).

titioning of the total nutrient return to the soil is given in Table 7.6, which illustrates the importance of rainwash. Although the main biomass nutrient store is in the structural tissue of trees, the major proportion of the annual return of nutrients to the soil is through leaf fall. The latter is of course continuous, while timber fall is infrequent.

The differences between the annual return of nutrients to the litter store (Table 7.5) and the measured instantaneous values for total nutrients in litter (Table 7.1) illustrate the rapidity of its decomposition. Sanchez (1973) estimates annual rates of litter decomposition to vary between 50 and 500 percent of the yield of litter measured at any one time.

TABLE 7.6
Components of nutrient gain from above in forest

	Nutrients (kg ha^{-1})				
	N	P	K	Ca	Mg
Rainwash	13	3.7	219	29	18
Timber fall	36	2.9	6	82	8
Leaf fall	199	7.3	68	206	45
Total	248	13.9	293	317	71

SOURCE: Nye and Greenland (1960).

■ 7.2.6 *Nutrient return from vegetation to soil in savannah.* The scale at which nutrients are returned to the litter layer in savannah vegetation is much smaller than in forest, and depends greatly on whether the grass has been burned or grazed. Under a regime that maintains a more or less stable community, we can assume that the return is roughly equivalent to the maximum aboveground nutrient store, since there is no long-term accumulation of tissue and most of the leaf and stem material produced in one year will be returned to the soil within 12 to 18 months.

■ 7.2.7 *The decomposition of organic matter.* The rate of breakdown of soil organic matter can be assessed in terms of the *decomposition constant*, or percentage of total organic matter mineralized each year. The main nutrients of concern here are nitrogen, sulfur, and phosphorus. In wet lowland forest, the decomposition constant is of the order of 2 to 5 percent per year, but in savannah in lower rainfall zones it may be as little as 0.5 to 1.5 percent, since the moist conditions necessary for decomposition occur for a shorter period each year.

The mineralization of nitrogen has already been discussed in 4.2.2. The factors affecting the rate of mineralization of phosphorus from organic matter appear to be similar to those for nitrogen (Birch, 1961), and organic phosphorus is mineralized at about the same rate as that of the organic matter fraction as a whole.

■ 7.2.8 *Nutrient uptake by forest and savannah.* What measurements have been made suggest that the root distribution with depth of well-developed savannah is similar to that of small forest trees, though the total root mass is lower. The mass and depth of rooting of tall forest communities appear to be greater. It has been estimated by Nye and Greenland that the annual uptake of nutrients from below 30 cm in a mature high forest is 40 N, 3 P, 58 K, 62 Ca and 12 Mg kg ha^{-1}. Thus over a fallow phase of, say, 10 to 20 years the cycling of nutrients up from the subsoil can be very substantial.

Again, the capacity of savannah vegetation to bring up nutrients from the subsoil, though less than that of high forest, will be greatly dependent on its treatment. Regular defoliation by fire or grazing will bring about a reduction in root mass and depth of rooting, and while nutrient cycling from depth in a protected dense savannah may be comparable to that of a light forest, in grassland that is continuously grazed by ruminants cycling is likely to be restricted to the upper soil layers.

■ 7.2.9 *Nutrient losses from leaching and runoff.* Under forest and savannah nutrient losses from leaching and runoff are small. In a closed community, whether tree- or grass-dominant, permanent root systems are likely to intercept nutrients otherwise liable to leaching, and surface erosion will be minimal. Table 7.7 shows the limited runoff and soil loss from forest compared with cropland and bare soil at four locations in West Africa differing in rainfall and slope. Nutrient loss from runoff has also been estimated on a subcontinental scale: it has been calculated from the silt load of the Amazon that the catchment as a whole, mostly forest, loses annually 4 K, 27 Ca, 5 Mg, 6 S kg ha^{-1}, and negligible amounts of N and P.

■ 7.2.10 *Nutrient gains from rainfall and nitrogen fixation.* The increment of nutrients from rainfall, if indeed they can be regarded as an increment and not merely as a redistribution, are small in relation to the total nutrient turnover in forest (see section 4.2.5), though they may be appreciable in relation to the turnover in savannah. Symbiotic legume nitrogen fixation is probably of significance in tropical forest,

TABLE 7.7

Runoff and erosion from forest, grassland and bare soil in W. Africa

				Runoff %			Soil loss (t ha^{-1})		
Locality	Period	% Slope	Rain (mm)	Forest	Crop	Bare Soil	Forest	Crop	Bare Soil
Ougadougou (U. Volta)	1967 to 1970	0.5	850	2.5	2–32	40–60	0.1	0.6–0.8	10–20
Sefa (Senegal)	1954 to 1970	12.0	1 300	1.0	21.2	39.5	0.2	7.3	21.3
Bouake (Ivory Coast)	1960 to 1970	4.0	1 200	0.3	0.1–26	15–30	0.1–0.2	0.1–26	18–30
Abidjan (Ivory Coast)	1950 to 1970	7.0	2 100	0.1	0.5–20	38	0.03	0.1–90	108–170

SOURCE: Charreau (1972).

where a number of the tree species are legumes, but the herbaceous legume component in tropical savannah is generally low. As indicated in section 4.2.4, we are now aware of nitrogen fixation in tropical grasses, but its quantitative significance in accumulating nitrogen under field conditions is not yet known.

Nonsymbiotic nitrogen-fixing bacteria are known in tropical forest and savannah soils, and it is generally believed that they contribute substantially, in forest at least, to the total accumulation of nitrogen in the fallow phase (Nye and Greenland, 1960).

■ 7.3 Biogeochemical and other changes on clearing and burning

■ 7.3.1 *Gain or loss of nutrients.* Because of the difficulties of sampling, there are few accurate figures for the gain of nutrients by surface soil layers when vegetation is cut and burned. The efficiency of the burn is, of course, a major determinant. Further, the distribution of ash is unlikely to be uniform; concentrations will occur where big logs are burned and rain after burning may redistribute ash by surface water movement.

When vegetation is burned, nearly all the nitrogen and sulfur is lost to the atmosphere as oxides, while phosphorus, potassium, calcium, and magnesium remain in the ash. Surface litter may also be consumed. Data from a 40-year-old forest in Ghana (Nye and Greenland, 1960) show the following changes in surface soil nutrients:

N:	minus 100 kg ha^{-1}	Ca:	plus 1,410 kg ha^{-1}
P:	plus 23 kg ha^{-1}	Mg:	plus 170 kg ha^{-1}
K:	plus 670 kg ha^{-1}		

The figures for P, K, Ca, and Mg corresponded roughly with the original data from the vegetation and litter store. Clearly in *citamene* systems, where additional wood is cut and brought onto the cleared area, the local increase in nutrients would be correspondingly greater. When savannah vegetation is burned, the increase in nutrient supply from ash is likely to be much lower than in forest, for obvious reasons.

■ 7.3.2 *Secondary effects.* The intense heat to which the surface soil is locally subjected when a large vegetation mass is burned on it has beneficial secondary effects. The soil is partially sterilized, with a resultant temporary modification to the microflora, leading to increased mineralization. There may also be a rise in pH as a consequence of the increase in bases. In addition, the seeds of annuals that might prove to be weeds in the ensuing crops will be killed. Obviously the greater the vegetation biomass, the fiercer the fire and the more substantial the effects. They are thus likely to be of more significance in forest than in savannah.

■ 7.4 The cropping phase

■ 7.4.1 Shifting cultivation practices bring about swift and drastic changes in the ecological situation: a closed and relatively stable vegetation is replaced by bare ground, which may be cultivated, and on which a crop community of fairly low biomass composed largely of annuals is grown. The biological and physical consequences of this transformation may in time include a decline in nutrient status of the soil and in its physical condition, an increase in arable crop weeds and grasses, and an increase in pests and diseases specific to the crops planted.

All these factors act toward a decline in crop yield with successive years of cultivation after the fallow vegetation, though this is not always the case between the first and second years of cropping. In situations where the initial clearing is incomplete, the better seedbed conditions of the second cropping year may mitigate the adverse effects of continued cropping. In addition, in some dense savannahs where the level of carbonaceous organic matter in the topsoil is high, nitrogen immobilization in the first year of cropping may lead to lower yields than in the second year (Vine, 1968).

There have been many surveys demonstrating the pattern of declining yield in forest environments with successive years of cropping: some examples are given in Figure 7.2. In addition, there have been numerous attempts to determine, by superimposing fertilizer treatments, which nutrients first become limiting when the cropping break is extended. The results of these tests vary widely with soil, fallow, and crop type, as would be expected. If one has to leap from a large body of variable data to a major generalization, it appears that in cropping after a forest fallow, available phosphorus is most frequently the primary limiting nutrient, and in cropping after savannah nitrogen is most frequently the primary limiting nutrient.

In the following brief review of ecological changes during cropping attention is confined to the pattern of change within a single cropping break. The long-term consequences of successive cycles of crop and fallow are considered in section 7.7.

■ 7.4.2 *Biogeochemical changes in the soil.* When forest land is cleared and burned, there is a sudden and substantial change in the soil environment. Burning in itself will consume most of the litter layer, and whether or not the rate of decomposition of organic matter increases after clearing—a possible consequence of cultivation, better aeration, more extreme wetting and drying, and higher soil temperatures—the rate of return of plant material to the soil surface is greatly reduced, at least

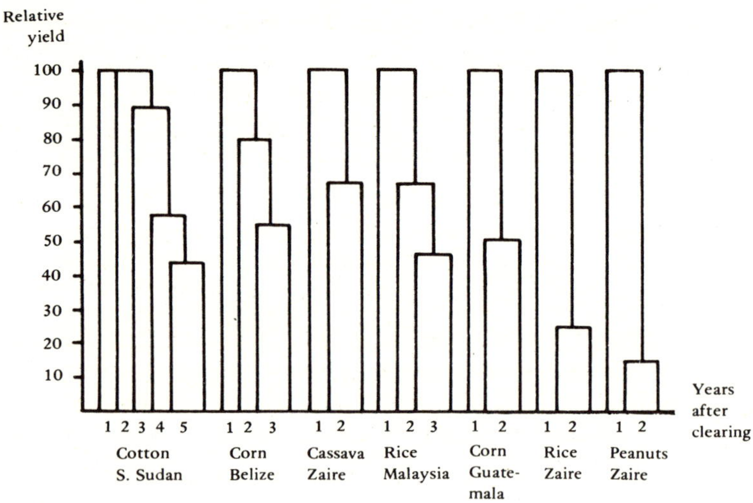

Figure 7.2. Decline in yield with continued cropping under shifting cultivation in forest environments (Ruthenberg, 1971).

relative to forest leaf fall. The result is a decline in soil organic matter content. In savannah, the rate of return of material to the soil may not differ so much between crop and fallow phases as in forest. On the other hand, the change from stable grassland to the open cultivated condition will almost certainly result in an increased rate of organic matter decomposition with the same ultimate effect—a decline in topsoil organic matter content.

The augmentation of available phosphate through burning may, in soils of high fixation capacity, be quite temporary, and so may be the rise in pH. Nye and Greenland (1960) and Mouttapa (1974) give data on the changes in chemical composition of soils with extended cropping after a fallow phase, but there have been few studies of the short-term changes within a single cropping break in actual shifting cultivation systems. Popenoe (1959), describing a situation in Guatemala, is an exception.

■ 7.4.3 *Leaching.* A second major ecological change on clearing and burning is the replacement of the permanent and often massive (in forest) root system of the fallow by a relatively weak and shallow annual or biennial crop root system that is active in intercepting nutrients for only a part of the year. A consequence of this is an increase in the potential for nutrients, particularly nitrogen and potassium, to be leached beyond the reach of crop roots. This will contribute to declining yield with extended cropping but, as we shall see later, it may not represent a complete loss to the cultivation system as a whole.

■ 7.4.4 *Erosion.* A further important ecological change is the sudden exposure of the topsoil to high-intensity rain. The soil surface, hitherto protected by a dense permanent canopy (except in regularly burned savannahs, the surfaces of which are exposed in early wet season until the canopy is reestablished), is protected in the cropping phase only when the crop or mixture of crops attains a complete canopy cover, which may be never. The consequences of this in relation to erosion are, naturally, greater on sloping sites. In addition, declining organic matter content with cropping will reduce the retention of water by the surface soil. However, though crop yield in shifting cultivation may be reduced in any one year by the chance occurrence of catastrophic high-intensity rain, the general level of nutrient loss through soil erosion is unlikely to contribute greatly to the decline in yield within a single short cropping phase.

■ 7.4.5 *Weed invasion.* The fourth major ecological transformation on clearing and burning is in the radiation environment near the ground and the degree of competition for water and nutrients in the surface soil. In forest, the surface soil is thoroughly permeated by tree and shrub roots and radiation levels near the ground are low. In savannah, light competition near the ground is intense and the surface soil is fully exploited by grass roots. These conditions militate against the establishment of arable weeds, which, having evolved in the open-field environment, are normally light-demanding and weakly competitive in soil already occupied by an active root mass.

Hence at the start of a cropping phase arable weed density tends to be low, and viable seed populations are at a low level as a consequence of burning. However, with extended cropping weeds may increase rapidly. Although it is difficult to separate the factors of declining available soil nutrients from increased incidence of weeds, it is probable that in many instances abandonment of the cropfield to fallow is triggered by the difficulties of controlling weeds. Invasion is particularly serious in savannah shifting cultivation: annual grasses, freed from competition from perennial grasses, increase rapidly in the cropfield situation; aggressive perennials, either incompletely destroyed by cultivation or advancing from the edges of the clearing, soon follow.

■ 7.4.6 *Pest and disease buildup.* Except in shifting cultivation systems where the crop "mix" is highly complex, the cropping phase is characterized by a concentrated assembly of a small range of genotypes. With repeated cropping of the same or related species, the probability of an increase in specific pests and diseases is self-evident, although in practice poorly documented.

■ 7.5 AGRONOMIC TACTICS IN THE CROPPING PHASE

■ 7.5.1 Various tactics have evolved to accommodate to the decline in fertility, the increasing incidence of weeds, and the recovery of woody vegetation as the cropping phase is extended.

It is normal for the important subsistence energy crops to be planted in the first year when growing conditions are at their best, as a security measure. Pride of place is also usually given to cash crops when they are grown, e.g., cotton in Africa. However, in some West African savannahs, nitrogen demanding crops are not planted in the first year because of limited availability of nitrogen. Thus in the Guinea savannah zone, yams (less demanding of nitrogen than grains) and legumes are planted in the first year and grain crops in the second year (Vine, 1968).

In the final year of the cropping phase, it is common to plant crops that exhibit one or more of the following characteristics:

a. tall crops able to compete with weeds and recovering fallow vegetation;
b. crops capable of yielding moderately well at low soil fertility levels;
c. crops with a long growing season able to utilize the abandoned land to the full, including biennials or perennials;

d. crops that are not in prime demand for subsistence but which constitute a food reserve.

Cassava, for example, has all these characteristics and is a common final crop; plantains or bananas and pigeon pea, the latter independent for nitrogen, satisfy some of the criteria. The farmer will often plant the final crop with the minimum of labor input and leave it to fend for itself while he concentrates on his newer and more productive cropfields. He may not even harvest it if his food supply from elsewhere is adequate.

■ 7.5.2 Table 7.8 shows the observed occurrence of some common crops in the first and last years of shifting cultivation systems in Zaire. The predominance of cassava as a final crop is clear. In forest regions the plantain tends to be a primary source of energy and is planted early; in the drier savannah grain crops are the primary source and plantains, when grown, may be planted last. Another characteristic that often distinguishes the late from the early stages of the cropping phase is a diminution in the range of crops planted in a single field, a reflection of the narrowing options imposed by the deteriorating environment.

TABLE 7.8
Occurrence of crops in the first and final years of the cropping phase in shifting cultivation systems. Zaire

Crop	Forest fallow		Savannah fallow	
	First crop	Last crop	First crop	Last crop
Bananas/plantains	44.1	16.7	0	2.7
Millets/sorghum	—	—	16.7	7.7
Corn	11.8	8.3	3.8	0
Cassava	8.4	41.7	3.2	36.2
Peanuts	—	—	14.3	5.0
Legumes	—	—	10.0	14.5
Yams	5.9	8.3	0	1.9
Sweet potatoes	—	—	7.4	5.0

SOURCE: Miracle (1967).
(Figures are percentages of number of observations made. Only the figures for first and last occurrence in one type of fallow for one crop may be compared.)

■ 7.6 THE RECOVERY PHASE OF FALLOW VEGETATION

■ 7.6.1 The most striking feature of the early recovery phase of a forest fallow after being cropped and abandoned is the rapid rate of accumulation of nutrients in the vegetation store. Table 7.9 shows the average annual rate of nutrient accumulation over different periods in an African forest (Nye and Greenland, 1960).

TABLE 7.9
Annual rates of nutrient accumulation in an African forest

	N	P	K	Ca + Mg
First 5 years	114	6	91	84 kg ha^{-1}
First 17 years	39	6	34	46 "
First 40 years	40	3	17	60 "

The reasons for high early rates of accumulation, particularly of nitrogen, have given rise to speculation. It is thought that uptake of nutrients from deep soil layers is particularly effective in early years (even though root systems may not be so well developed as in more mature forest) because the tree roots are tapping a store of nutrients leached from the crop root zone during the arable phase. The above high early rates of uptake of nitrogen and potassium, nutrients likely to have been leached, would support this view. Some indication of the subsequent progressive accumulation of nutrients in the vegetation store may be gathered from Table 7.1, which compares secondary forests of different ages from 5 to 40 years.

■ 7.6.2 There is very little information on the rate of accumulation of nutrients by savannah during recovery from cropping. Much would depend on whether it was burned or grazed during this phase. We may speculate, however, that with protection from grazing and fire, savannah nutrient accumulation would rapidly achieve a relatively steady state condition, since the build-up to a closed grassland would be more rapid than the succession to a mature forest. The only factor acting progressively to increase nutrient accumulation would be the gradual replacement of annuals by deeper-rooted perennials as the succession advanced.

It also seems fairly certain from experimental work that the changes in surface soil structure between the savannah fallow and the cropping phase are relatively transient. Structure is re-created when the land is abandoned, but on recultivation structural effects do not extend beyond one or two years of cropping (Pereira et al., 1954).

■ 7.6.3 In some shifting cultivation systems, the process of nitrogen accumulation during the recovery phase is consciously hastened by the planting of nitrogen-fixing species at the end of the cropping phase. Thus in Papua–New Guinea, casuarina is planted on abandoned sweet potato fields and the timber subsequently used for fencing, etc. On the Thailand-Burma border, the Karen and Lahu peoples plant pigeon pea (here a short-lived perennial) in upland rice stubble before abandoning the field; the grain is harvested for stock.

■ 7.6.4 In the former Belgian Congo, now Zaire, the colonial administration instituted a regulated system of shifting cultivation, known as the *couloir* or corridor system. A land area was laid out in 100 m strips, each family being allotted about 10 ha in a block across the strips. The cultivators moved annually across a series of about 10 strips, along their individual blocks, cropping 1 to 2 ha yr^{-1}. To assist in recolonization of the abandoned strip, the "cultivation front" did not move laterally one strip a year, but leap-frogged one strip each year and returned later to the uncleared strips. In this way the strip of abandoned plots was flanked by regenerating forest on both sides. The method is described by Ruthenberg (1971). Apart from stabilizing the cropping pattern, such a degree of organization helped communications and technical extension. However, the system has since been abandoned.

■ 7.7 Long-term effects of shifting cultivation

■ 7.7.1 Although the adverse effects of a single short cropping break on the chemical status and physical condition of the soil under forest and savannah may be relatively small, these effects become cumulative if the fallow phase is regularly too short to be fully restorative. We can expect a steady decline in organic matter content which will in time stabilize at a lower level than when the vegetation was first cleared for cropping. The level at which this stability is reached will bear a general relation to crop yield and to the productivity of the farming system as a whole.

■ 7.7.2 There are a number of factors involved in this regression. First, if the duration of the fallow under a cropping cycle of a given cultivation frequency is inadequate for the primary community to reestablish itself, the subclimax vegetation will not be as effective in restoring soil organic matter content, fertility status, and physical state. As well as reducing crop yield in the arable phase, the ecological recovery succession in the fallow phase is further retarded. This cycle is progressive: primary forest is replaced by secondary forest, secondary forest by grassy scrub, grassy scrub by grass savannah. With declining biomass and rooting depth, the restorative capacity of the fallow per unit time is steadily reduced.

Second, soil changes operating to reduce the yield potential of the crop–fallow cycle may also be autogenic. Loss of topsoil organic matter may reduce infiltration rate, thereby increasing runoff and further loss of topsoil by erosion. Periodic cropping may induce surface soil impermeability through physical changes. Soil and vegetation changes interact progressively: a steady decline in fertility status and physical condition will further degrade the vegetation succession.

Third, the overutilization of a given area for cropping may force the cultivator on to land less suited to long-term use. Fourth, cultivation techniques become more drastic: as forest gives way to savannah, the digging stick gives way to the hoe, planting on the flat gives way to mounding systems, etc. These techniques have an accelerating effect on organic matter decomposition. Finally, declining fertility levels may enforce a change in the crops grown to those less desirable or nutritious but of lower fertility demand, e.g., from yams and corn to cassava.

■ 7.7.3 We can distinguish, in the progression from shifting cultivation to semi-intensive cropping, cultivator responses that may be characterized as negative or positive. Shortening of the fallow phase, use of less suitable land, increasingly drastic cultivation, a switch to more exhaustive crops—all are negative responses in the sense that their implementation merely hastens the decline in productivity to which they are a reaction. Positive responses are those that are truly restorative: the return of animal and vegetation residues to the soil, soil conservation practices, the use of mineral fertilizer, etc.

Under conservative shifting cultivation systems the fertility levels of the primary forest community, which can take more than a century to reestablish itself, may never be regained, but stability is attained at a satisfactory level. In West Africa, under a long period of 2-year crop and 10-year fallow, soil organic matter content stabilized at about 75 percent of that of primary forest (Reed, 1951). Thus, shifting cultivation is not an inherently exploitative system. However, such situations, in a world of rapidly increasing population, are becoming rarer and rarer. At the opposite pole, in regions characteristic of much of India, where a dense population has been developing over a long period, the level of productivity at which stability has been reached under what by now are continuous cultivation systems is very low indeed. In between are vast areas where progressive intensification consequent upon population increase, unaccompanied in the majority of instances by restorative measures, is creating increasing problems of environmental stability and food supply.

REFERENCES

Birch, H. F. 1961. Phosphorus transformations during plant decomposition. *Plant and Soil* 15:347.

Bourlière, F., and Hadley, M. 1970. The ecology of tropical savannahs. *Annual Review of Ecology and Systematics* 1:125.

Charreau, C. 1972. Problèmes posés par l'utilisation agricole des sols tropicaux par des cultures annuelles. *Agronomie Tropicale.* Série Agronomie Générale. Etudes Techniques no. 9. Paris: IRAT.

Miracle, M. P. 1967. *Agriculture in the Congo basin.* Madison: University of Wisconsin Press.

Moss, R. P. 1969. The ecological background to land use studies in tropical Africa, with special reference to the West. In: Thomas, M. F., and Whittington, G. W., eds. *Environment and land use in Africa.* London: Methuen.

Mouttapa, F. 1974. Soil aspects in the practice of shifting cultivation in Africa and the need for a common approach to soil and land resources evaluation. In: *Shifting cultivation and soil conservation in Africa.* FAO Soils Bulletin no. 24. Rome.

Nye, P. H., and Greenland, D. J. 1960. *The soil under shifting cultivation.* Harpenden, England: Commonwealth Bureau of Soils. Technical Communication no. 51.

Pereira, H. C., Chenery, E. M., and Mills, W. R. 1954. The transient effects of grasses on the structure of tropical soils. *Empire Journal of Experimental Agriculture* 22:148.

Popenoe, H. 1959. The influence of the shifting cultivation cycle on soil properties in Central America. *Proceedings of the 9th Pacific Science Congress* 7:72.

Reed, W. E. 1951. *Reconnaissance soil survey of Liberia.* U. S. Department of Agriculture. Agricultural Information Bulletin no. 66. Washington, D. C.

Ruthenberg, H. 1971. *Farming systems in the tropics.* Oxford: Clarendon Press.

Sanchez, P. 1973. Soil management under shifting cultivation. In: *A review of soils research in tropical Latin America.* North Carolina Agricultural Experiment Station Technical Bulletin no. 219.

Vine, H. 1968. Developments in the study of shifting agriculture in tropical Africa. In: Moss, R. P., ed. *The soil resources of tropical Africa.* London: Cambridge University Press.

Chapter 8

Shifting Cultivation Systems: Hydrological and Energetic Aspects

■ 8.1 HYDROLOGICAL ASPECTS

■ 8.1.1 In considering the hydrological aspects of shifting cultivation systems, we must distinguish between the small-scale phenomena associated with a single cropfield over a single cropping phase and large-scale phenomena associated with shifting cultivation as a continuing practice within a catchment area over a long period. The hydrological cycle within a cropfield in a shifting cultivation system does not differ in principle from that of a cropfield in a semi-intensive or intensive rainfed cropping system, which will be dealt with in chapter 12. Furthermore, the biogeochemical consequences of runoff and erosion in shifting cultivation have already been examined in the previous chapter. There are, however, one or two special hydrological characteristics of shifting cultivation systems, which are considered below.

■ 8.1.2 *Small-scale effects.* A common feature of shifting cultivation, particularly in areas where the better classes of land have been developed for more intensive cropping, is that it is carried out on steep slopes. In this era of rapidly increasing population, it is a familiar experience when flying over tropical mountain regions to note the gradual extension of cropping up hillsides from a well-cultivated valley. The contrast is particularly clear in geologically young regions of sharp relief: in Southeast Asia, for example, there is often an abrupt boundary between intensive stable wet rice farming on narrow flat valley floors and destructive shifting cultivation on the forested slopes immediately above. (It is not always destructive, however: the Karen people on the Burma-Thailand border have for long operated a dual system of permanent valley wet rice and shifting hill rice, which, through careful and conservative cultivation techniques, has remained very stable.) A recent general account of the progressive deterioration of mountain environments with human occupation is given by Eckholm (1975).

Just what effect slope will have on water gain or loss by run-on and run-off water will, of course, depend on local conditions. On very steep slopes in high-rainfall areas runoff and erosion may be catastrophic when the whole hillside slumps down following a period of intensive rain. It is relatively rare to find any water control or conservation structures in shifting cultivation fields. Where crops are planted in rows and not irregularly, the rows tend to run up and down the slope rather than across, probably for ease of weeding. Occasionally, unburned logs are placed across the slope to prevent soil movement, and in the New Guinea Highlands quite elaborate systems of drainage have evolved for sweet potato cropping, but these are the exceptions rather than the rule.

The amount of runoff and the efficiency of water use by crops are also influenced by cropping tactics. In section 6.6, the complex cropping mix characteristic of shifting cultivation systems in lowland high-rainfall regions was mentioned; the greater the complexity of cropping in time and space the greater the degree of protection of the soil surface from erosion and the more efficient the use of soil water by crops. In the simpler types of cropping pattern associated with lower-rainfall regions, the same degree of surface protection and crop water use may only be achieved during the short period when crops reach their maximum leaf area. In summer-rainfall climates, there is a vulnerable period between the start of the wet season and the time of attainment of a full crop canopy, during which the soil surface is exposed to rain.

By careful organization of cropping in space and time, limitations to soil water supply may be minimized or overcome. A simple but elegant example is seen in the corn-opium poppy systems of the H'mong people in the North Thailand hills, who plant corn early in the wet season and opium poppies in the corn stubble late in the wet season, the opium poppies maturing largely on stored water. Normally, the opium poppy area, requiring heavy thinning, weeding, and harvesting inputs, is smaller than the corn area. The subsequent opium poppy crop is often grown in the lower central part of the corn area, where soil water is available for the longest period after the rains have ended.

■ 8.1.3 *Large-scale effects.* The magnitude of large-scale and long-term effects of a shifting cultivation system operating over time within a specific catchment area will depend upon the stability of the vegetation-soil complex. This in turn will be governed by the basic hydrological characteristics of the catchment, the cultivation frequency of the cropping system, and the capacity of the vegetation to recover during the fallow phase.

Primary forest is hydrologically stable in that the amplitude of runoff and drainage fluctuations and the amount of erosion are at a minimum for the particular conditions of soil, slope, and rainfall. However, even if the cropping system imposed on it were of a cultivation frequency low enough to ensure a virtually complete return to the original hydrological condition—a situation becoming less and less common in the world today—the capacity of the fallow to absorb rainfall during the early period of recovery could well be below that of the primary forest.

If cultivation frequency increases to the point where major successional changes take place—degradation to secondary forest, to scrub, and to grass savannah—the hydrological properties of the catchment are likely to decline further. Although during the growing season the water-absorbing capacity of a dense stand of savannah grassland may be as good as that of secondary forest, when the grassland is burned regularly in the dry season or heavily grazed the soil surface is exposed, particularly in early wet season. Furthermore, the shallower soil layer exploited by the root systems of degraded communities is likely to amplify the fluctuations of drainage. Pereira (1973) gives a useful appreciation of the topic.

These changes not only influence in the long run the productivity of the cropping system, but more importantly, because of the greater number of people affected, may adversely affect the rainfall/runoff characteristics of the catchment as a whole and influence settlement and agriculture downstream. Peak wet season flow is increased, which can lead to flooding, and the quantity and duration of dry-season flow is reduced, which can affect irrigation schemes and domestic water supplies. The potential for damage is greatest in Southeast Asia, where the basic hydrological pattern is that of intensive wet rice cultivation in valley floors and deltas fed by streams draining forested catchments of sharp relief in zones of high rainfall.

■ 8.2 Energetic aspects: general characteristics of energy inputs

■ 8.2.1 *Simplicity.* The energy input-output balance of shifting cultivation systems is usually fairly simple, not by virtue of the fact that they are shifting, but because they are largely subsistence systems, because little or no fertilizer is used, because the maintenance of livestock is not a common major enterprise, and because forms of power input other than human labor are virtually unknown. The equation is thus often reduced to a simple pattern of human effort in, human food out. In this section on inputs, *gross* rates of work energy expenditure are quoted.

■ 8.2.2 *Advantages over more intensive systems.* It has already been mentioned that at least in climates with a distinct dry season, shifting cultivation systems have an advantage over intensive rainfed annual cropping systems in that total energy inputs for cropping are spread over a longer period: the wet season inputs of planting, weeding, and harvesting are balanced by the dry season inputs of clearing and burning. It is misleading to regard the clearing and burning of fallow vegetation as an additional burden on the shifting cultivator compared with the farmer's input in more intensive cropping systems, for the operations are the shifting cultivator's means of providing adequate nutrients for his crops. The more intensive cultivator, to compensate for the lack of a periodic accession of available nutrients through clearing and burning, has to expend energy in carting manure to his fields or has to purchase energy in the form of mineral fertilizer. Alternatively, he may do without nutrient inputs, which implies either a lower crop output or an increase in energy input in cultivating a larger area to compensate for lower output per hectare. The "hidden" energy cost of the nutrients released by clearing and burning may be equated with the energy stored in the vegetation that is burned, and is analogous to the indirect energy cost of manufactured fertilizer. Accepting a figure of 20 MJ kg^{-1} for the energy value of forest timber (Lieth, 1975), a 20-year-old secondary forest with a standing dry matter yield of 100 t ha^{-1} would be storing approximately 2×10^6 MJ ha^{-1} above ground, which would be lost as heat on burning.

■ 8.2.3 *Disadvantages in comparison with intensive systems.* In other ways, however, the character of shifting cultivation systems may be associated with higher energy costs than more intensive rainfed cropping systems. The fact that cropfields are generally scattered and at some distance from the domicile implies a substantial expenditure of energy in walking, often with loads of produce, tools, etc. In a study in Zaire, the average distance from house to fields was 1.5 km, and within a farm family of four persons an average of 4 to 5 journeys was made each day. Such an effort would require an energy expenditure of about 6 MJ day^{-1} or approximately one-sixth of an estimated total family expenditure of 35 to 40 MJ day^{-1}. In New Guinea, in a system that is examined in more detail later (Rappaport, 1971), cartage amounted to 20 percent of the total work energy input.

In addition, shifting cultivation is often practiced in hilly and remote localities. Steep topography clearly involves additional energy expenditure in walking, and remoteness imposes extra effort in movement to shops to purchase essential supplies and to markets if saleable crops are grown. The heavy inputs required to market bulky produce are illustrated in a later example.

■ 8.2.4 *Forest and savannah.* It might be assumed *a priori* that preparing land for cropping in forest requires a higher energy input than preparing land in savannah. This does not seem to be the case. Although the effort required to cut down trees and to heap them for burning is greater than that required to slash grass, savannah cultivation normally involves a greater input for cultivation and seedbed preparation, which more than compensates for the difference in energy costs of clearing. Seedbed preparation inputs are particularly high when laborious techniques such as composting of grass, mounding, bed-making, or ridging are employed.

Table 8.1 shows the energy input into clearing and cultivation for five locations differing in fallow vegetation. Slashing and felling inputs were lower in savannah than in forest, but the efforts required for burning, clearing up, and cultivation were higher. The high figures for rainforest in the Congo are the result of very careful clearing, involving two burning operations.

■ 8.3 Energetic aspects: analysis of inputs

■ 8.3.1 In order to gain an appreciation of the range of energy inputs in the wide range of shifting cultivation systems three examples, of increasing complexity, are analyzed:

a. A subsistence cropping system without domestic animals: the Iban of Sarawak, growing upland rice as their main energy food.

b. A subsistence cropping system with domestic meat animals: the Tsembaga of New Guinea, growing sweet potatoes, taro, and other noncereal energy foods and raising pigs.

c. A subsistence-cash crop system in Zaire, where the farmers grow rice, cassava, and bananas and market a proportion of all three crops.

■ 8.3.2 *Subsistence cropping without domestic animals.* The Iban people of Sarawak in Southeast Asia cultivate upland rice in high-rainfall forest and also grow a range of minor crops. Cultivation frequency is low—about 5 to 10 percent—the population at the time this study was made (Freeman, 1955) being fairly sparse. In fact, it is not uncommon for primary forest to be cleared and cropped. The Iban live in family groups, or *bileks*, averaging 5 to 6 persons; the groups live in communities of 4 to 50 *bileks*. In good years, the average yield of paddy rice is 2 to 2.7 t ha^{-1}, in poor years about 0.7 to 1.0 t ha^{-1}.

TABLE 8.1
Gross energy input into land clearing in shifting cultivation systems (MJ ha^{-1})

Operation	Kalimantan, primary forest	Kalimantan, secondary forest	Guatemala, forest	Zaire, rainforest	Zaire, savannah	Zaire, *Pennisetum* savannah
Slashing and felling	307–342	154–189	290	328–517	233	44
Burning and clearing	34–86	34–86	158	182–1,006	737–927	819
Total	341–428	188–275	448	510–1,523	970–1,160	863

SOURCE: Adapted from data collated by Ruthenberg (1971); man-hours converted to MJ assuming 1.05 MJ hr^{-1}.
NOTE: Savannah clearing includes hoe cultivation.

Table 8.2 shows the inputs for cultivation tasks in primary and secondary forest. (The division of labor tasks between males and females is worth noting.) Although, as expected, clearing inputs are higher for primary forest, weeding inputs are lower, reflecting the slow ingress of arable weeds on land that has never been cropped. In all, there is little difference in energy requirement for cropping between the two types of forest. The energy required to transport the crop to storage sites is very small; presumably the Iban live near their cropfields. Furthermore, as will be seen in the two following examples, transport energy for a crop high in dry matter such as rice grain is less than that for fresh root crops such as cassava or sweet potatoes. Total energy input varies from 553 to 790 MJ ha^{-1}, which is low, reflecting the rather rough clearing effort and the lack of any need for seedbed preparation.

■ 8.3.3 *Subsistence cropping plus livestock.* The most detailed published study of the energetics of shifting cultivators is that of Rappaport (1971), who investigated the farming system of the Tsembaga people in the highlands of New Guinea. The Tsembaga grow a range of crops, the main energy foods being sweet potatoes, taro (both *Colocasia* and *Xanthosoma*), yams, bananas, cassava, and sugarcane. Sweet potatoes and cassava are used for pigfeed as well as for human consumption. Cultivation frequency in these high-altitude forests and savannahs is about 10 percent according to Rappaport, but this is not typical of the New Guinea highlands as a whole, which in places are densely populated (Brookfield and Brown, 1963). The Tsembaga normally crop in two localities, in the higher fields growing mainly sweet potatoes for pigs and in the lower fields concentrating more on crops

TABLE 8.2
Gross energy input in shifting cultivation
(a) The Iban in Sarawak (MJ ha^{-1})

Operation	Sex of operator	Primary forest	Secondary forest
Slashing	M + F	67	55–67
Felling	M	133–156	44–56
Clearing	M	22–56	22–56
Dibbling	M	44	44
Sowing	F	56	56
Weeding	F	133–178	166–222
Harvest	F	155–222	155–222
Transport	M	11	11
Total		621–790	553–734

SOURCE: Dabasi-Schweng (1974); original data from Freeman (1955); converted from man-days ac^{-1} assuming 5 hr day^{-1} and 0.9 MJ hr^{-1}.

for human consumption. There is little crop storage, and food requirements are harvested more or less continuously every one or two days.

Table 8.3 shows the distribution of energy input between the various field tasks. The biggest single input is that of planting and weeding, making up almost one-third of the total, a proportion of the same order as in the previous example. Here, however, the resemblance ends, for where the Iban clear the forest only roughly and do very little cultivation, land clearance and seedbed preparation by the Tsembaga is a very thorough operation and also includes fencing the cropfield against pigs and placing logs along the contour to prevent soil erosion.

TABLE 8.3
Gross energy input in shifting cultivation
(b) The Tsembaga in the New Guinea Highlands (MJ ha^{-1})

Operation	Energy expenditure
Clearing underbrush	585
Clearing trees	235
Fencing garden	353
Weeding and burning	195
Placing soil-retainers, etc.	151
Planting and weeding	1,867
Other maintenance	477
Sweet potato harvest	464
Taro harvest	59
Cassava harvest	22
Yam harvest	160
Cartage	1,240
Total	5,808

SOURCE: Rappaport (1971).

The heavy input for cartage has already been noted; in this instance bulky root crops high in water content are being moved. The system is clearly high-input in character, total energy expenditure exceeding 5,000 MJ ha^{-1}. As will be seen later, a substantial proportion of this effort is devoted to the production of pigfeed.

■ 8.3.4 *Subsistence and cash cropping.* Table 8.4 shows a similar partitioning of production tasks in a rice-cassava-banana shifting cultivation system in the Congo (Ruthenberg, 1971; original data Tondeur, 1956). Cultivation frequency is 10 percent: 2 years cropping and 18 years forest fallow. One hectare is cropped and half a hectare is cleared each year, and a proportion of the output of all three crops is marketed.

TABLE 8.4
Gross energy input in shifting cultivation
(c) in Zaire (MJ ha^{-1})

Operation	Energy expenditure
Clearing and burning	336
Seeding and planting	126
Rice harvest (incl. transport to hut)	189
Preparation of rice	21
Transport of rice to market	22
Cassava harvest (incl. transport to hut)	210
Preparation and drying of cassava	504
Transport of cassava to market	134
Banana harvest (incl. transport to hut)	67
Transport of bananas to market	266
Total	1,875

SOURCE: Ruthenberg (1971); original data from Tondeur (1956); converted from man-days ha^{-1} assuming 4 hr day^{-1}, 1.05 MJ hr^{-1} for work other than transport, 0.80 MJ hr^{-1} for transport.

Total energy input is fairly high, nearly 2,000 MJ ha^{-1}, but nearly one-quarter of this is expended in transporting the cash crop portion to market. In addition, these data include processing of rice and cassava both for home consumption and for sale, and in neither of the two previous examples was processing mentioned. If processing and transport to market are excluded to make the data comparable, total energy input is reduced to 928 MJ ha^{-1}.

In sharp contrast to the New Guinea example, the input for weeding was apparently so small that the investigator did not record it. It may be assumed that in this fairly wide crop-fallow sequence in rainforest, weed infestation is light and seedbed preparation and cultivation inputs are low. On the other hand, the example brings home forcibly the importance of the processing component and the effort required to market bulky produce.

■ 8.3.5 *Comparison of different ages of fallow.* In Table 8.5 an instructive example is presented of the differences in input for the same type of forest shifting cultivation system between fallows of varying age. The data are from the lowlands of Guatemala (Carter, 1969) in a twin-peak high-rainfall climate, where the Kekchi people grow a wide range of crops—Carter lists harvesting operations for 24 of them—of which corn is the main energy staple.

In "mature" forest over eight years old, felling trees is a major input, but the effort required for weeding is negligible. In younger forest, still recovering from the previous cropping phase, all vegetation can be slashed down by machete, which reduces clearing inputs substantially. On the other hand, the work input into weeding increases rapidly as the fallow period is reduced. Carter's data are very complete, and it is worth noting that the operations of site selection, granary building, carting and storing the crop, and preparing it for consumption, often

TABLE 8.5
Gross energy input in shifting cultivation
(d) Comparison of three ages of forest fallow; Guatemala (MJ ha^{-1})

Operation	Fallow 8 yrs	Fallow 4–7 yrs	Fallow 1–3 yrs
1. Site selection	5	3	3
2. Slashing	71	119	88
3. Felling	315	—	—
4. Firebreaking	10	27	27
5. Burning	5	2	2
6. Planting corn	46	46	46
7. Fencing	5	5	5
8. Watching	23	23	23
9. Weeding	2	59	110
10. Intercropping	108	92	44
11. Harvest	104	118	102
12. Building granary	24	24	24
13. Cartage	86	103	82
14. Storing	42	46	42
15. Shelling corn	8	9	8
Total (production tasks*)	689	491	447
Total	854	676	606

SOURCE: Carter (1969); original data in man-hours, converted at: 0.5 MJ hr^{-1} for operations 1, 8, 15; 0.8 MJ hr^{-1} for operations 13, 14; and 1.05 MJ hr^{-1} for all other operations.
*Excluding operations 1, 12, 13, 14, 15.

not included in "production task" data, amount on average to 24 percent of the total input.

■ 8.4 ENERGETIC ASPECTS: ANALYSIS OF OUTPUTS

■ 8.4.1 *Subsistence cropping without domestic animals.* Output data for four Iban families growing upland rice in Sarawak are given in Table 8.6. In comparison with the averages given by Freeman (1955) (see section 8.3.2), yields per hectare are low. Because of this, one family produced only three-quarters of what it needed; for two others production and requirement were roughly in balance, and one family produced a substantial surplus.

Freeman's estimate of annual rice requirement per head, 314 kg, may appear to be high. However, if we make the assumptions set out

TABLE 8.6
Energy output in shifting cultivation. The Iban of Sarawak

	Family 1	Family 2	Family 3	Family 4
Persons per family	8	6	5	5
Labor force (over 15 yr)	4	4	4	3
Area of rice (ha)	2.63	2.20	1.92	2.02
Total rice yield (kg)*	1,950	1,760	1,710	2,640
Yield per ha (kg)	741	800	891	1,360
Yield as % of normal family requirement**	77.8	93.4	108.7	168.4

SOURCE: Dabasi-Schweng (1974); original data from Freeman (1955).
* Assuming 2.75 kg per "ganteng."
**Estimated at 314 kg $head^{-1}$.

below, the correspondences with Freeman's estimates are remarkably close:

a. Each worker has an energy requirement equivalent to that of an adult male: 10 MJ day^{-1}.
b. Each nonworker has half this requirement.
c. Milling percentage of rice 60.
d. Energy value of milled rice 15 MJ kg^{-1}.

On the basis of the above, the estimates of yield as a percentage of family requirement come out as 80.1, 86.7, 106.7, and 162.7, respectively, for the four families (compare with Table 8.6).

The average value for gross energy input from Table 8.2, taking the mean of upper and lower values for both types of forest, is 664 MJ ha^{-1} for production tasks only. If we accept the mean yield of paddy rice from Table 8.6 as output, 934 kg ha^{-1}, and the assumptions of rice milling percentage and energy value as above, consumable output is 8,410 MJ and the output/input energy ratio based on net work energy expenditure (0.75 of gross) is about 17:1.

Clearly, producing upland rice under shifting cultivation in Sarawak is energetically fairly efficient. However, Table 8.6 shows that for families with a high proportion of nonworkers and in years of below-average yield the contribution of the staple rice crop to total energy needs is inadequate. Presumably Family One supplemented its diet with other crops and with the products of hunting and fishing, or purchased rice, or went hungry.

■ 8.4.2 *Subsistence cropping plus livestock.* For the Tsembaga of the highlands of New Guinea, although energy input per hectare is much higher than that of the Iban, total food energy output is proportionately greater (Rappaport, 1971). Rappaport estimated total food energy output to be 95,400 MJ ha^{-1}, which with a gross input of 5,800 MJ ha^{-1} yields an output/input ratio of 16:1. Assuming a ratio of net to gross work energy expenditure of 0.75, this becomes 22:1, somewhat better than that of the Iban. Furthermore, of the total food energy output, humans consumed only about three-fifths and the remainder was fed to pigs. The raising of a substantial surplus of pigs is a cultural obligation for the Tsembaga, who accumulate a herd over a number of years for a periodic festival at which most adult and many young pigs are killed and eaten. It is estimated by Rappaport that apart from food energy, each pig requires a human energy input of 188 MJ yr^{-1} for management, and he quotes an example of a herd size of 169 pigs at the peak of a "pig festival cycle."

Thus Tsembaga agriculture is essentially a high input–high output system, of fairly high efficiency in respect of the ratio of consumable energy output to production task input. It is cushioned from seasonal yield variation by the animal production component, in that only three-fifths of the total food energy output is needed for direct consumption; in fact, the pigs are raised on a scale that in strict economic terms could be regarded as surplus to requirement. However, pigs also provide protein to supplement the protein-deficient root and tuber crops that make up the overwhelming proportion of the vegetable diet. In this context it is interesting to note that some New Guinea highlanders living largely on a sweet potato diet appear to have evolved an intestinal flora that includes nitrogen-fixing bacteria (Ooman, 1970).

■ 8.4.3 *Subsistence and cash cropping.* In the data on the Zaire system, which was a "model" estimate of Tondeur (1956) and may thus have been optimistic in outlook, it was assumed that half the rice, half the bananas, and all the cassava were sold. If we accept the yield estimates given and make realistic assumptions of the consumable energy value of the crop products, it would appear that the output/input ratio based on net energy expenditure in production tasks only is 50–

60:1, or 25–30:1 if processing and transport to market are taken into account. This figure of 50–60:1 for output/input ratio, though suspiciously high in comparison with the general level for rainfed hoe cultivation systems (Black, 1971), receives some support from the findings of Chandra et al. (1974) on the energetic efficiency of root and tuber crop production. The proportion of consumable energy output retained for farm use is of the order of 20 percent. However, if we assume a farm family of five to seven persons, this would amount to little more than half of their total energy requirements, and a figure of about 40 percent retained would seem more realistic.

Nevertheless, though the model estimates of the proportions of crop output consumed or sold may be sanguine, it is clear that the system, requiring only a moderate energy input per hectare, is efficient in the sense of providing a substantial energy surplus. This efficiency is clearly associated with the forest environment, a low cultivation frequency and the production of root and tuber crops; except from the effort involved in clearing, food crops can be raised in the short arable phase with the minimum of input.

■ 8.4.4 *Energy inputs other than human energy.* In the estimates of output/input ratio made above, human energy constituted the whole of the input side. As already mentioned, draft animals are rare in shifting cultivation and mechanical power even rarer. Some fuel, e.g., kerosene, may be purchased for domestic use, but the only purchased input for crop production is likely to be the indirect energy cost of hand tools amortized over their life; a negligible amount. In the example immediately above, purchased tools amounted in monetary terms to less than one percent of gross returns.

REFERENCES

Black, J. N. 1971. Energy relations in crop production—a preliminary survey. *Annals of Applied Biology* 67:272.

Brookfield, H. C., and Brown, P. 1963. *Struggle for land.* Melbourne: Oxford University Press.

Carter, W. E. 1969. *New lands and old traditions. Kekchi cultivators in the Guatemalan lowlands.* Center for Latin American Studies. Gainesville, Florida: University of Florida Press.

Chandra, S., de Boer, A. J., and Evenson, J. P. 1974. Economics and energetics; Sigatoka Valley, Fiji. *World Crops* 26:34.

Dabasi-Schweng, L. 1974. Economic aspects of shifting cultivation. In: *Shifting cultivation and soil conservation in Africa.* FAO Soils Bulletin no. 24. Rome.

Eckholm, E. P. 1975. The deterioration of mountain environments. *Science* 189: 764.

Freeman, J. D. 1955. *Iban agriculture.* Colonial Research Studies. London: H. M. Stationery Office.

Lieth, H. 1975. Measurement of caloric values. In: Lieth, H., and Whittaker, R. H., eds. *Primary productivity in the biosphere.* New York, Heidelberg, Berlin: Springer-Verlag.

Ooman, H. A. P. C. 1970. Interrelationship of the human intestinal flora and protein utilization. Joint Symposium with the Royal Society of Tropical Medicine and Hygiene. Nutritional problems of the tropics. *Proceedings of the Nutrition Society* 29:197.

Pereira, H. C. 1973. *Land use and water resources in temperate and tropical climates.* London: Cambridge University Press.

Rappaport, R. A. 1971. The flow of energy in an agricultural society. *Scientific American* 224:117.

Ruthenberg, H. 1971. *Farming systems in the tropics.* Oxford: Clarendon Press.

Tondeur, M. G. 1956. L'agriculture nomade au Congo Belge. FAO. *Agriculture Nomade* 1:15.

Chapter 9

Semi-Intensive and Intensive Rainfed Annual Cropping Systems: General Aspects

■ 9.1 Introduction

■ 9.1.1 The range of cultivation frequency from shifting cultivation systems to those rainfed cropping systems where the arable area of the farm is cropped every year is continuous; the subdivisions of shifting, semi-intensive and intensive cultivation are a taxonomic convenience only. As indicated in the first chapter, Ruthenberg's (1971) demarcation of these classes by cultivation frequencies of 30 and 70 percent respectively is being followed.

In section 6.1.2, reference was made to the nomenclature of the less intensive rainfed cropping systems. The terms "bush fallow" (Boserup, 1965) or "rotational bush fallow" (Morgan, 1969) characterize the shorter fallow phases of semi-intensive cultivation but are often applied to the system as a whole. The adjective "rotational" reflects the approach toward a regular crop-fallow time sequence as cultivation frequency increases.

■ 9.1.2 There is one arbitrary element to be introduced into the definition of "rainfed." This group of chapters will be concerned with *upland* rainfed cropping only, that is, nonirrigated cropping under nonflooded conditions, and not with rainfed wet rice-growing.

There are three basic systems under which rice is grown:

a. *upland or hill rice*, where the crop is grown in a nonflooded situation, often on quite steep slopes, like any other rainfed cereal;
b. *rainfed wet rice*, where the crop is grown in a flooded situation on natural rainfall plus natural run-on water;

c. *irrigated wet rice*, where the crop is grown in a flooded situation where water supply and drainage are controlled.

(By far the bigger proportion of tropical wet rice is not irrigated; for example, in Thailand, the second largest exporter of rice in the world, over three-fourths of the rice area is rainfed.)

In this book it is convenient to discuss rainfed wet rice under irrigated cropping systems, since the energetic, hydrological, and biogeochemical aspects of the system approach those of irrigated wet rice more nearly than they approach upland rice.

■ 9.1.3 Semi-intensive rainfed cropping systems are most characteristic of tropical Africa, and within Africa of the western regions. In tropical America, semi-intensive systems are found in areas where population increase has enforced a narrowing of the crop-fallow sequence: this situation is more characteristic of summer-rainfall zones and mountain regions than of the lowland wet tropics. In Southeast Asia the contrast between intensive lowland cropping and shifting cultivation on uplands remains relatively sharp, and semi-intensive systems are not prominent.

Intensive rainfed cropping systems, where the arable area of the farm unit is cropped virtually every year, are most characteristic of the densely populated lowland areas of South and Southeast Asia, particularly the Indian subcontinent and Java. In Africa, the regions of dense population and intensive cropping are more localized: for example, North Nigeria. The heavily exploited Andosols of the high Andean valleys are a typical American example of intensive rainfed cropping.

■ 9.2 TERMINOLOGY OF COMPLEX CROPPING PATTERNS

■ 9.2.1 In our earlier consideration of shifting cultivation systems, reference was made to various aspects of what we may collectively term "complex cropping," an omnibus phrase used here to describe any situation where more than one crop is growing on a given land area at any one time.

The breadth of the term is appropriate to shifting cultivation systems, which in many instances are characterized by a wide range of crops grown in a complex time and space pattern. However, as will be seen later, the transition from shifting cultivation to more intensive systems is usually associated with the emergence of simpler and more readily recognizable cropping patterns within a single season. In order to be in a position to discuss these, some definition of the basic types is needed. With increased research interest in intensification of cropping, not only in the tropics but elsewhere, a terminology is emerging, though it has not yet stabilized. Andrews and Kassam (1976) quote a series of definitions contributed by P. A. Sanchez, which differ slightly from those given below.

In this book, the classes of complex cropping are defined as:

a. *Mixed cropping.* More than one crop on a given land area at one time, but not arranged in a geometric pattern.

b. *Intercropping.* More than one crop on a given land area at one time arranged in a geometric pattern. For example, two rows of sorghum alternating with two rows of peanuts.

c. *Relay cropping.* More than one crop on a given land area planted at different times, the later crop (or relay crop) being planted before the earlier crop or crops are harvested.

Where more than one crop is grown in a year on the same land but the crops follow each other and do not overlap, the term *sequential cropping* is appropriate. The more general phrase *multiple cropping* is often used fairly loosely to signify any extension of the cropping season beyond the span of a single crop, whether this is achieved by relay or sequential planting.

■ 9.2.2 However, the above definitions are subject to conventional interpretation. Thus intercropping, although in strict terms subsuming relay cropping, is usually reserved for situations where two (or more) crops are planted at the same time. Furthermore, relay cropping is generally applied only to situations where the crops are arranged in a geometric pattern, though strictly it is applicable also to nongeometric mixed cropping situations where crops are planted at different times. Finally, the term mixed cropping has two connotations: it refers to crop mixtures where each species or cultivar is planted separately (often in its particular ecological niche) and also to situations where a mixture of seed, usually a range of cultivars within a species, is sown together. The former pattern is characteristic of complex shifting cultivation systems, the latter of some cereal cropping systems in Africa. Rainfed multiple cropping practices and patterns in Asia, America, and Africa are reviewed by Harwood and Price (1976), Pinchinat et al. (1976) and Okigbo and Greenland (1976) respectively.

■ 9.3 Biological, physical, and socioeconomic transitions

■ 9.3.1 Intensification of cropping and population increase are strongly associated, in the tropics and elsewhere. This is not the place to discuss philosophical issues in geography, but it is worth mentioning that the identities of the dependent and the independent variables in this relationship have been the focus of controversy. Boserup (1965) is perhaps the protagonist most closely associated with the view that intensification of cropping is a consequence of population increase rather than the reverse, and her hypothesis appears to be prevailing.

■ 9.3.2 The broad biological, physical, and socioeconomic characteristics of an increase in rural population in a given tropical region are many and varied. The most important of these are summarized in Table 9.1. Population increase is associated with a change from shifting cultivation to semi-intensive to intensive cropping, involving an increase in length of the crop phase and a decrease in the fallow phase. On the socioeconomic side, it is associated with an increase in the importance of cash cropping relative to subsistence, with a clarification and legalization of land tenure, and with a change in social organization toward more permanent nucleated settlements in closer touch with each other and with the world in general. Gleave and White (1969) discuss the relation between population density and agricultural systems in West Africa in some depth.

■ 9.3.3 The primary biological and physical effect of increasing human pressure on land exploited under shifting cultivation systems is a progressive decline, through intensification, in soil fertility and hence crop productivity. This regression has already been examined in section 7.7 and the topic will be treated more fully in chapter 11. At this stage all we need to note is that in the regression in productivity that occurs when the restorative capacity of the fallow phase is unequal to the demands made upon it by the cropping phase, one of three situations obtains:

a. Soil fertility and crop productivity decline steadily toward some relatively stable level at which the farming system is capable of sustaining itself, though at a low level of output.
b. Stability in fertility status and crop productivity is not reached and the condition of the farming system represents some point in a deteriorating environmental cline.

TABLE 9.1

Characteristics and components for a model of African population/land relationships

POPULATION	LOW DENSITY	→ *Increasing numbers*	→ HIGH DENSITY
			Possible surplus rural population
		Seasonal/Permanent migration	→ Urban drift
SYSTEM	SHIFTING CULTIVATION	ROTATIONAL CULTIVATION FALLOW	SEMI-PERMANENT/ PERMANENT CULTIVATION
	→ *Increasing length of cultivation period*	→ *Decreasing length of fallow period*	→ *Manuring and fertilizing*
CROPS	SUBSISTENCE FOOD CROPS	→ *Decreasing importance*	→
	CASH (FOOD AND EXPORT) CROPS	→ *Increasing importance*	→
TENURE	COMMUNAL RIGHTS TO LAND (individual usufructary rights)	→ *Communal rights decreasing individual rights increasing*	→ INDIVIDUAL RIGHTS TO LAND
	Land allocation by need	→ Land transfer by pledge, rent, lease and sale	
	Fragmented/dispersed holdings	→ Consolidated holdings	
	No permanent demarcation of holdings	→ Permanent demarcation of holdings	
SETTLEMENT	IMPERMANENT/ MIGRATORY SMALL VILLAGES/ DISPERSED	→ *Increasing permanence and nucleation*	→ PERMANENT/FIXED NUCLEATED AND DISPERSED
EXCHANGE	NONEXISTENT/LOCAL	→ *Increasing involvement at local, regional, national and international levels*	

SOURCE: After Protheroe (1972).

c. The decline in fertility status and crop productivity is arrested or reversed by fresh technology: e.g., the application of manure or fertilizer, the implementation of soil conservation measures, etc.

Apart from changes in soil fertility and in restorative methods, we can classify the main transitions from shifting cultivation to intensive rainfed cropping under the following heads:

a. Changes in field and cropping pattern.
b. Changes in cultivation methods, tools and power inputs.
c. Changes in the livestock element.

These are considered in order beginning with section 9.4.

■ 9.3.4 To conclude the more general part of this chapter, mention must be made of the problem of discussing change. Apart from reversions associated with war, pestilence, or access to new land by colonization, the history of world agriculture is one of intensification. Thus at any one location it is a one-way temporal process that may be treated historically. On the other hand, farming systems of varying degrees of intensification exist simultaneously. Since this book is concerned with tropical farming systems as they are currently represented in the world, and does not purport to be a history, the synchronic approach is appropriate. However, as other authors have clearly discovered, it is often more felicitous to write of the differences in degree of intensification in historical terms, as changes in time at one place rather than as differences between places or systems at one time. The reader is warned that such a descriptive mode, where employed, is merely a semantic convention to facilitate the comparison of contemporaneous systems.

■ 9.4 Changes in Field and Cropping Pattern

■ 9.4.1 The changes in field and cropping pattern in the progression from shifting cultivation to intensive rainfed cropping may be grouped into the following five categories:

a. Consolidation of cropfields.
b. Systematization of the crop-fallow sequence.
c. Systematization of cultivation frequency between fields.
d. Specialization in crop location between fields.

e. Simplification and systematization of the cropping pattern within a field.

■ 9.4.2 *Consolidation of cropfields.* In shifting cultivation systems, the individual cropfields within a single farming unit are irregularly scattered as islands surrounded by fallow-phase land. (This is not always the case: in the highly organized Lua system on the Thai-Burma border, where upland rice is grown on a 1-year crop, 10-year fallow system, groups of farmers may clear the whole of one hillside in one year and abandon it in the following year for another hillside.) As the proportion of fallow to crop land declines, the boundaries of the individual crop and short-fallow areas tend to become contiguous and permanently defined. On the other hand, long-term, dense occupation of an area by a rural population may in time lead to fragmentation of holdings as a result of inheritance processes, as we have seen.

■ 9.4.3 *Systematization of the crop-fallow sequence.* With an increase in cultivation frequency is associated a crystallization of the crop-fallow sequence into the semblance of a "rotation" which, in a more intensive phase, becomes a rotation of crops only.

There have been attempts, as with shifting cultivation sequences, to classify the types of sequence encountered in semi-intensive farming systems. Ruthenberg (1971) quotes a fairly complex taxonomy from von Rotenhan (1968). The variability is so wide and continuous that perhaps only three concepts can be recognized:

a. *The type sequence.* This would represent the mean of the basic pattern for a number of like farming units measured over a period of years. It might constitute a more or less regular alternation of say, 2 to 7 years of cropping with 3 to 4 years of fallow, giving a range of cultivation frequencies from 30 to 70 percent.
b. *Long fallows.* A moderately regular crop–fallow sequence on a given land area may be broken by a longer fallow. If land area permits, the cultivator may retire a section of his farm to fallow for a substantial period if it is clear that the sector is depleted of fertility, subject to erosion, etc.
c. *Opportunistic fallowing.* In reality, the basic sequence is modified from year to year according to available labor, the current number of consumers, the carry-over of stored food, or the current incentive for cash cropping.

Partial systematization of the crop–fallow sequence is a natural outcome of the increase in ratio of crop land to fallow land, and represents a reduction in the options available to the farmer in choosing what land he will cultivate and what he will leave to fallow.

■ 9.4.4 *Systematization of cultivation frequency between fields.* With intensification and consolidation of the holding, differences in the frequency with which any one area within the farm is cultivated tend to become crystallized. The concentric zonation of cropping frequency around the domicile has already been noted, the higher frequency nearer home often being sustained by a higher level of manuring. Such a zonation, however, is largely associated with a dispersed settlement pattern of isolated homesteads. Where the farming community is nucleated into villages, the main cropping areas may be at some distance from the house. In this situation, a clear distinction tends to emerge between the fields, on which a small number of crops are grown in a simple cropping pattern, and the home garden, where a range of fruit, vegetable, and household crops are grown in a complex pattern. The latter is likely to receive most if not all of the animal manure available and will be largely devoted to subsistence crops, whereas if cash crops are grown they are likely to be in the outlying fields. Such patterns are described for Southeast Asia, where they are highly developed, by Terra (1958).

■ 9.4.5 *Specialization in crop location between fields.* In semi-intensive and intensive cropping systems operating within a defined and limited farm area, the cultivator must necessarily locate his crop so as to optimize the relations between crop requirement and land capability. Thus fertility-demanding crops will predominate on fertile bottom land, crops requiring free drainage on elevated land, etc. This specialization is to some degree a contrast with the approach of the shifting cultivator, who on the whole selects land for clearing and cropping on the basis of its general capability to produce, and then exploits the variability within a single cropfield by a carefully located mixture of crops.

■ 9.4.6 *Simplification and systematization of the cropping pattern within a field.* The change from shifting cultivation to intensive cropping is usually accompanied by a reduction in the number of crops grown in each field at one time, by a more regular pattern in the mix of crops, and by the adoption of geometric planting. The principles governing the sequence of crops within a single cropping phase in shifting cultivation systems that have already been noted are likely to hold good for semi-intensive cropping systems.

The reduction in the number of crops grown in each field at one time is clearly related both to the localization of specific crops (see section 9.4.5) and to the development of a greater distinction between cropfields and home gardens (see section 9.4.4). In addition, intensification is associated with the emergence of fairly standardized intercropping and relay cropping practices, replacing the less orderly planting and harvesting procedures associated with shifting cultivation. This change may be to some extent related to a decline in the marginal productivity of labor with intensification, enforcing a more systematic approach to work input.

Finally, there appears to be a general relation between cultivation frequency and crop geometry, in that crops in more intensive systems are more likely to be sown in rows. It can be readily appreciated that if only one or two crops are being grown in the field there is no reason not to plant them in a regular fashion. Row planting, however, has no necessary association with the use of draft animals, for in the great majority of annual cropping systems they are used only for land preparation and not for subsequent intercrop operations.

■ 9.4.7 *Plant populations.* In advanced temperate agriculture, it is normal for farmers to attain, or to attempt to attain, crop population densities close to the optima recommended through research via extension. In contrast, one of the most common observations in rainfed tropical farming regions is that crop populations are suboptimal by experiment station standards. There are four possible reasons for this:

a. With inadequate extension, the farmer may be unaware of the optimum crop population. This is the least likely cause; traditional technology is usually firmly based on known crop behavior.
b. He may sow or plant enough seed or vegetative material for a full stand, but poor seedbeds, low-quality planting material, or subsequent pest, disease, or weed incidence may reduce it.
c. He may, consciously or instinctively, be aiming to maximize yield per unit labor rather than per unit area; e.g., low crop populations may make weeding easier even though weed yield is higher.
d. He may be aware of the population that gives the highest

yield in average or good seasons, which will be the one recommended, but deliberately adopts a population that is optimal for the below-average situation, since he would prefer to be sure of less than maximum yield rather than risk greater loss by having too high a population in an adverse season.

■ 9.5 Changes in cultivation methods, tools and power inputs

■ 9.5.1 With increasing cultivation frequency and crystallization of field boundaries, there is a steady progression in the cropfield towards fully cleared land, partially an attrition due to repeated cultivation but often the result of a deliberate effort to increase cropping efficiency. Fully cleared land is of course essential where draft animals are used. That the conversion of land formerly under shifting hoe cultivation to the fully cleared condition is no mean task for the nonmechanized farming unit is illustrated by an example of residual obstructions on formerly forested land in Northeast Thailand (Haswell, 1973):

Standing trees ha^{-1}	13
Stumps ha^{-1}	37
Termite mounds ha^{-1}	5
Total obstructions ha^{-1}	55
Average farm size (ha)	2.67
Obstructions per farm	147

■ 9.5.2 The types of hand tool used in shifting cultivation have already been described. Once the cropping system has become semi-intensive, the friable surface soil characteristic of the burned forest disappears, the necessity to break up grass root systems increases, and arable weed infestation becomes a greater problem. In addition, in short-fallow systems cultivation tends to become more thorough to encourage the decomposition of carbonaceous organic matter. All these changes demand tools more substantial than the digging stick for planting or the machete or cutlass for weed-slashing, and the hoe becomes universal.

■ 9.5.3 The above changes are merely a progression in hand tools. The major leap forward in cultivation power associated with intensification is the use of draft animals and their modern successors, the small tractor or hired large tractor.

The use of cattle and buffalo as draft animals (rarely horses or mules, which were military or transport animals) is of great antiquity in Asia, and their possession is, broadly speaking, a function of farm size and general affluence. In intensive cropping systems, both rainfed and irrigated, the Asian farmer may maintain no herbivorous livestock other than a pair of draft animals. In tropical Africa, the plough and draft animals were only introduced about 150 years ago, and hoe cultivation is still widespread. There is a recognizable distinction in Africa between pastoral peoples who maintain cattle and practice little cropping and agricultural peoples who farm with the hoe and do not own cattle either for draft or other reasons. In some instances, the draft animal phase is being leapfrogged in a direct advance from hand tools to mechanization. In tropical America, the plough and draft cattle are post-Columbian and a high proportion of crop farmers still use the hoe, even if at the same time they maintain cattle.

■ 9.5.4 This is not the place to embark on a consideration of the mechanization of tropical farming systems, except to note that its initial impact has been greatest with respect to stationary motors for tasks such as pumping, milling, etc., rather than as a source of motive power. Where tractors are found, they tend to be associated with the larger farms and with high input–high output irrigation cropping systems. The general economic status of the small rainfed farm rarely permits the ownership of tractors, though they may be hired.

■ 9.6 Changes in the livestock element

■ 9.6.1 The role of livestock in tropical cropping systems will be accorded a chapter to itself, and at this point attention will be concentrated on changes that occur when cultivation frequency increases. However, in order to appreciate these changes, it is necessary to emphasize that the rationales for maintaining livestock in tropical cropping systems extend well beyond the immediately obvious ones of production for sale, production for subsistence, or for draft.

We can summarize the roles of livestock as:

a. *Production role.* For sale or subsistence.
b. *Investment role.* As a current financial buffer, or an insurance against old age or incapacity.

c. *Sociocultural role.* As an element of social prestige and influence, or to fulfill cultural obligations.
d. *Energy role.* To provide draft power or fuel in the form of dung.
e. *Nutrient role.* To provide manure.

The topic is considered in more detail in chapter 17.

■ 9.6.2 As we have seen, the livestock component of shifting cultivation systems, where it exists, commonly takes the form of nonruminants, pigs and poultry, which may forage on free range but to which a proportion of the carbohydrate crop output and perhaps protein is diverted. Where ruminant livestock are maintained in shifting agriculture, they are not used for draft nor to any great extent to provide manure. Their role is normally that of an investment or an element of sociocultural influence. There is little or no integration between the cropping element and the ruminant livestock element, which subsists almost exclusively on free range grazing.

In semi-intensive cropping systems, the place of nonruminants as converters of crop carbohydrate to meat and egg protein is maintained; in addition the crop farmer may own small herds of cattle, sheep, or goats, largely as an investment but partially for meat and milk. In regions characterized by semi-intensive cropping the proportion of nonarable land is generally adequate to support ruminant herds, usually under communal grazing, with only a small labor input.

With further intensification, the role of ruminant livestock for draft and as a source of manure increases in importance. Furthermore, if the region is not characterized by a substantial proportion of inherently nonarable land, the area of fallow and wasteland on which stock can graze freely becomes limiting. This involves a higher input of labor: to herd cattle to outlying areas, to cut natural forage and bring it home, or to conserve crop residues for feeding in periods of scarcity. The cultivation of forage crops for ruminants, whether draft or investment stock, is minimal in rainfed agriculture. The final intensification in the role of livestock, associated with prolonged dense human settlement and hence a shortage of wood, is the utilization of cattle and buffalo dung for domestic fuel, a widespread practice on the Indian subcontinent.

REFERENCES

Andrews, D. J., and Kassam, A. H. 1976. The importance of multiple cropping in increasing world food supplies. In: American Society of Agronomy. *Multiple cropping.* Madison, Wisconsin: ASA Special Publication no. 27.

Boserup, E. 1965. *The conditions of agricultural growth.* Chicago: Aldine.

Gleave, M. B., and White, H. P. 1969. Population density and agricultural systems in West Africa. In: Thomas, M. F., and Whittington, G. W., eds. *Environment and land use in Africa.* London: Methuen.

Harwood, R. R., and Price, E. C. 1976. Multiple cropping in tropical Asia. In: American Society of Agronomy. *Multiple cropping.* Madison, Wisconsin: ASA Special Publication no. 27.

Haswell, M. R. 1973. *Tropical farming economics.* London: Longman.

Morgan, W. B. 1969. Peasant agriculture in tropical Africa. In: Thomas, M. F., and Whittington, G. W., eds. *Environment and land use in Africa.* London: Methuen.

Okigbo, B. N., and Greenland, D. J. 1976. Intercropping systems in tropical Africa. In: American Society of Agronomy. *Multiple cropping.* Madison, Wisconsin: ASA Special Publication no. 27.

Pinchinat, A. M., Soria, J., and Bazan, R. 1976. Multiple cropping in tropical America. In: American Society of Agronomy. *Multiple cropping.* Madison, Wisconsin: ASA Special Publication no. 27.

Protheroe, R. M. 1972. *People and land in Africa south of the Sahara.* London: Oxford University Press.

von Rotenhan, D. 1968. Cotton farming in Sukumaland. In: Ruthenberg, H., ed. *Smallholder farming and small holder development in Tanzania.* Afrika-Studien no. 24. Munich: IFO Institut.

Ruthenberg, H. 1971. *Farming systems in the tropics.* Oxford: Clarendon Press.

Terra, G. J. A. 1958. Farm systems in south-east Asia. *Netherlands Journal of Agricultural Science* 6:157.

Chapter 10

Semi-Intensive and Intensive Rainfed Cropping Systems: Energetic Aspects

■ 10.1 Introduction

■ 10.1.1 It is unfortunate that the level of detail at which human energy inputs in crop production have been studied is somewhat lower for semi-intensive and intensive cropping systems than for shifting cultivation systems; on the other hand the number of data sets is larger. Anthropologists, with their interest in less advanced societies, have concentrated upon shifting cultivation systems. Their approach is normally more detailed than that of agricultural economists, to whom we are indebted for most of our information on more intensive cropping systems. Furthermore, agricultural economists concerned with the tropics have until recently been little interested in energy concepts, and have confined their attention to input/output relations in monetary terms, though it is heartening to see the beginnings of a joint approach in Chandra et al. (1976).

■ 10.1.2 In chapter 8, the energetics of contrasting shifting cultivation systems were examined in some detail. Thus in Tables 8.1, 8.2, 8.4, and 8.5 specific values for gross energy expenditure were assumed for different tasks and, in Table 8.2, for a particular estimated body weight. In keeping with the more general approach of this chapter, and of chapter 15, standard assumptions have been made concerning human and draft animal energy expenditure, which have been applied to data on hours or days of human or animal labor. The assumptions are based on Philipps (1954) and Brody (1968).

For human energy expenditure, they are:

Basal metabolism	0.25 MJ hr^{-1}
Gross work energy expenditure	1.00 MJ hr^{-1}
Net work energy expenditure	0.75 MJ hr^{-1}

The values are appropriate for a 55 kg farm worker working on average five hours a day. Gross work energy expenditure for both humans and draft animals varies from 3 to 8 times the basal metabolic rate (Brody, 1968); in this instance a factor of 4 has been assumed, on the low side of the range to allow for rest periods. With a five-hour working day, assuming eight hours of sleep at 0.25 MJ hr^{-1} and a gross energy expenditure when neither working nor sleeping of 0.4 MJ hr^{-1}, daily gross energy expenditure is 11.4 MJ. The fraction of this value assigned to gross work energy expenditure (which, as we shall see, is an important characteristic), is therefore 0.44. This is very close to the value of 0.43 obtained by Passmore and Durnin (1955) for British miners.

The assumptions for draft animals are:

Basal metabolism	1.5 MJ hr^{-1}
Gross work energy expenditure	6.0 MJ hr^{-1}
Net work energy expenditure	4.5 MJ hr^{-1}

The values are appropriate for a 550 kg draft animal working about six hours a day. They are in each instance 6 times the values assumed for humans, and hence the ratio of gross work energy expenditure to basal metabolism is again 4:1. With a six-hour working day, assuming a gross energy expenditure of 2.0 MJ hr^{-1} when not working (including grazing, ruminating and resting), daily gross energy expenditure is 72 MJ, close to the feed energy intake of 91 MJ day^{-1}, less 17 MJ day^{-1} in dung energy output, measured by Odend'hal (1972) for adult male cattle in India. The fraction assigned to gross work energy expenditure is therefore 0.5. Revelle (1976), in his assessment of energy use in rural India, assumed the fraction of 0.43 that Passmore and Durnin estimated for humans.

■ 10.2 General level of energy input and output

■ 10.2.1 The first comparative study of energy input and output in nonmechanized tropical cropping systems was that of Black

(1971), who used data on crop yields and man-hours of labor in crop production tasks from a number of sources. Black's findings for rainfed hoe cultivation systems are presented in Table 10.1. The first three examples are shifting cultivation systems.

In Table 10.1 the level of input varies widely, from only 71 MJ ha^{-1} for *Digitaria exilis*, a secondary cereal grown largely as a quick catch crop, to nearly 2,000 MJ ha^{-1} for upland rice in the Philippines. The range of output/input ratio is correspondingly wide, from 3 to 34,

TABLE 10.1
Energy input and output for rainfed hoe cultivation systems (a)

Country	Crop	Net energy input (MJ ha^{-1})	Energy output (MJ ha^{-1})	Output/ input ratio
Sarawak	Upland rice	1,047	22,710	22
Sarawak	Upland rice	318	10,827	34
Philippines	Upland rice	1,886	34,107	18
Ghana	Corn	411	10,249	25
Gambia	Millet	331	2,946	9
Gambia	Millet	402	3,654	9
Gambia	Sorghum	100	2,024	20
Gambia	*Digitaria*	71	1,320	19
Gambia	Corn	478	8,166	17
Gambia	Upland rice	1,253	3,884	3
Mexico	Corn	901	15,423	17
Mean		654	10,483	17.5 (unweighted)

SOURCE: Black (1971); original data in man-hrs ha^{-1}, assuming work energy expenditure at 0.63 MJ hr^{-1}.

giving an unweighted average of 17.5. If a net work energy expenditure rate of 0.75 MJ hr^{-1} had been assumed, the unweighted average output/input ratio would have been 14.7.

■ 10.2.2 Another series of examples, from rainfed cropping systems in Africa, is given in Table 10.2. The figures are adapted from a survey of data by Clark and Haswell (1970). Clark and Haswell used as a common measure of crop output the kilogram wheat equivalent, based on the market price of the several crop products relative to wheat. However, since they give these equivalents for the crops concerned, it is possible to reconvert their data and, by applying appropriate food energy values, to arrive at energy output/input ratios. Only data for rainfed cereal and noncereal energy crops have been extracted. The original table included also figures for grain legumes, but the value of legumes resides in their protein rather than their energy yield and calculation of energy output/input ratios has little meaning.

The unweighted average output/input ratio for cereals is 18.5, somewhat higher than that from Black's data computed on the common basis of 0.75 MJ hr^{-1} net work energy expenditure. The range, however, is almost equally wide. The average output/input ratio for noncereals is substantially higher, 29.5, and only one value is less than 20.

■ 10.2.3 *Comparison of cultivation frequency.* Although comparable sets of figures are limited, there is an indication from Table 10.1 and from the data presented in chapter 8 that human energy input

TABLE 10.2
Energy input and output for rainfed hoe cultivation systems (b)

Country	Crop	Net energy input (MJ ha^{-1})	Output/ input ratio
(a) CEREALS			
Ghana	Corn	488	16
U. Volta	Corn	312	23
Nigeria	Corn	229	24
Gambia	Millet	408	7
U. Volta	Millet	292	13
Uganda	Millet	615	8
Cameroon	Sorghum	525	22
E. Africa	Sorghum	341	39
Gambia	Sorghum	124	16
U. Volta	Sorghum	292	17
Mean		363	18.5 (unweighted)
(b) NON-CEREALS			
U. Volta	S. potato	439	21
E. Africa	Cassava	772	20
U. Volta	Cassava	195	54
Nigeria	Cassava	266	27
Ghana	Yams	1,249	26
U. Volta	Yams	443	9
Nigeria	Yams	814	44
Uganda	Banana	870	35
Mean		631	29.5 (unweighted)

SOURCE: Adapted from Clark and Haswell (1970); for method see text.

per hectare is somewhat higher for shifting cultivation than for more intensive cropping systems. If this is a valid generalization, then presumably the difference lies largely in the additional clearing and burning effort associated with shifting agriculture.

However, if in comparing shifting cultivation with intensive cropping we were to include the energy cost of the burned fallow vegetation (see section 8.2.2), then shifting cultivation appears to be grossly "wasteful" of energy. On the other hand, the appellation wasteful only has validity if there is some real alternative use for the vegetation that is burned. Furthermore, if the system is stable, in the sense that cultivation frequently is low enough to permit a return to a comparable level of vegetation energy accumulation at the end of each fallow phase, then the energy resource has been fully renewed by photosynthesis. The periodic burning can thus be regarded as an intelligent use of renewable energy resources to provide available nutrients for crop production. (The "waste" comes in the volatilization of nutrients, nitrogen and sulfur, on burning.)

There is insufficient evidence to demonstrate whether the apparently higher human energy input for shifting cultivation is associated with a higher average crop energy output (due to higher soil fertility status, fewer weeds, etc.), and it is risky to rule on the relative output/input ratios of shifting and more intensive systems. The same is true of the comparison between semi-intensive and intensive systems. Environmental variability and a relatively small sample restrict us to the very general conclusion that for tropical annual hoe-cultivation systems the output/input ratio for cereals is likely to be in the range of 10 to 20.

■ 10.2.4 *Comparison between crop types.* There appears to be a contrast in Table 10.2 between cereals and noncereal carbohydrate crops—cassava, bananas, etc. Admittedly the figures in Table 10.2 are scanty; nevertheless, the validity of the comparison is supported by data presented in chapter 8 and also by the work of Chandra et al. (1976) in Fiji. The output/input ratios they calculated are summarized in Table 10.3. The basis of their calculations is not given.

Chandra et al. compared farming systems at two locations, and hence differences between crops are confounded with crop production practices and the comparison thereby weakened. For example, at Naceva the generally lower values for output/input ratio are partially attributed to the use of tractors for land preparation, fuel for which augmented the human energy inputs. Even so, a trend toward higher output/input ratios for noncereal carbohydrate crops seems apparent, supporting the data presented in Table 10.2 and in chapter 8.

■ 10.3 Distribution of inputs according to production tasks

■ 10.3.1 It was stressed earlier that although more sets of data on energy input and output are available for semi-intensive and intensive cropping systems than for shifting cultivation, in general the level of

TABLE 10.3
Output/input ratios for cereal and noncereal crops. Fiji

Naceva		Naduri-Emuri	
Rice	20	Cassava	71
Corn	17	Corn	54
Irish potato	23	Sweet potato	82
		Yam	77
		Banana	130

detail is lower. The data presented in Table 10.4, which gives examples of the distribution of total crop production energy input between tasks, is of necessity reduced to three categories only: land preparation plus planting, weeding, and harvesting plus threshing. In some instances, a separation into four tasks has been possible.

■ 10.3.2 The main contrast between the relatively intensive rainfed cropping systems, data for which are given in Table 10.4, and shifting cultivation systems, shown in Tables 8.2–8.5, lies of course in the virtual elimination of inputs for clearing and burning as cultivation frequency approaches 100 percent. In the four examples of shifting cultivation shown in Tables 8.2, 8.3, 8.4, and 8.5, the proportions of total energy input for production tasks that can be clearly assigned to clearing and burning are 30, 22, 36 and 42 percent respectively.

On the other hand, as expected, this is compensated for by an increased input for land preparation and planting with increasing cultivation frequency. For the shifting cultivation systems illustrated in Tables 8.2, 8.4, and 8.5, land preparation and planting inputs amount to only 13 to 17, 13, and 8 percent respectively of total production task inputs, whereas Table 10.4 shows that for more intensive cropping systems the proportion averaged 29 percent.

■ 10.4 Output/input ratios and subsistence

■ 10.4.1 In this book the conventional method of expressing output/input ratio for a farming system or an individual crop enterprise has been adopted; that is, the total crop energy output relative to the total net energy expended in production tasks. It is also clear from chapter 8 that in calculating output/input ratios on this basis difficulties arise in assigning tasks to "production": the gray areas include, for example, processing and the transport of crops both within the farm and to market. However, this source of imprecision in estimating the energetic efficiency of crop production is of minor importance compared to the relation between output/input ratio as it is conventionally expressed and the true food energy balance of the subsistence farm family.

TABLE 10.4
Distribution of human energy input in semi-intensive and intensive rainfed cropping systems according to production tasks
(percent of total input)

Country	Crop	Land prep.		Plant-ing	Weed-ing	Harvesting and threshing
Senegal	Peanuts		22		50	28
U. Volta	Millet		18		38	44
U. Volta	Corn		32		54	14
Tanzania	Corn/ Sorghum		41		27	32
	Cotton		25		29	46
Senegal	Millet		11**		72	17
Cameroon	Sorghum	.25		22	39	14
Malagasy Republic	Cassava	34		16	17	33
Senegal*	Peanuts		30		24	46
Tanzania*	Cotton		8		46	46
Senegal*	Millet	29		4	38	29
Taiwan*	S. potato	2**		30	27	41
Mean			29		38	33

Source: Ruthenberg (1971).
* Oxplough systems; remainder hoe cultivation.
**Planted with little cultivation.

■ 10.4.2 It is perhaps identifying the obvious to point out that the farmer or farm worker is expending energy throughout the day and not merely while he is directly engaged in cultivating crops. He derives his daily energy needs largely from the food energy output of the farm unit; hence in relation to the net energy output of the unit, it is total gross energy expenditure that constitutes the real input denominator.

The average amount of time over a year spent by workers in production tasks varies according to the environment (particularly the length of growing season) and the type of crop, and is perhaps also influenced by ethnic factors. However, the weight of evidence (Clark and Haswell, 1970; Haswell, 1973) suggests that in tropical cropping systems the year-round average approaches four to five hours. If we assume an average gross rate of energy expenditure in crop production tasks of 1 MJ hr^{-1} and a total daily gross energy expenditure of 10 to 12 MJ (both for adult males), then four to five hours field work represents about three- to six-tenths of total gross energy expenditure.

However, the above refers only to those working, whereas in subsistence farming the unit is almost certainly supporting members of the family who are merely consumers. Though virtually everyone who is capable of working—wives, adult relatives, older children—may contribute in some degree to crop production, young members are consumers only and moreover there is generally an older generation to support. We cannot apply the conventional worker-to-consumer ratio in this instance, since we are dealing in energy terms. Neither is it man-hours of work divided by total man-hours, even when converted to adult male equivalents, since energy expenditure while working is greater than when not working. The appropriate ratio is that of the energy expended in crop production by the farm family relative to total energy expended through the day or year. It is very difficult to arrive at any average figure, but it is not unrealistic to assume that to make allowance for the non-working component of the farm family would reduce the ratio, gross energy expended in crop production to total gross energy expended, from the three- to six-tenths range for the individual worker to something of the order of two- to four-tenths.

Furthermore—and this is particularly true of low-technology rainfed cropping systems and the more arid climates—crop yields vary widely from season to season. Environmental conditions are unfavorable and facilities commonly inadequate for any major carryover of stored food from one year to the next. Moreover, even when the quantity of crop output minus storage and processing losses is theoretically adequate if rationed out over the year, there is a natural tendency to eat well after one harvest and to suffer "seasonal hunger" before the next.

■ 10.4.3 The situation described above is formalized in Table 10.5 for ranges of output-to-input ratio as normally calculated (10 to 20), for the ratio of gross energy expended in crop production to total gross energy expended (0.2 to 0.4), and for negative departures from average crop yield (1 to 0.5). It is suggested that an "ultimate" output/input ratio of 1.5 or less represents a critical situation in view of the general magnitude of storage, processing, and household waste losses. That such situations are not reached as frequently as the table might suggest is perhaps largely due to the buffering effect of food from minor

TABLE 10.5
Energy balance of subsistence farm families (food energy output/gross family energy expenditure) as influenced by workers' output/input ratio, by ratio of crop production energy expenditure to total energy expenditure, and by below-average crop energy output

Workers' output/ input* ratio	Ratio of gross crop production energy expenditure to total gross energy expenditure by farm family	Negative departures from average crop energy output (ratio average/ actual)		
		1.0	0.75	0.5
20	0.4	6.0	4.5	3.0
	0.3	4.5	3.4	2.2
	0.2	3.0	2.2	1.5
15	0.4	4.5	3.4	2.2
	0.3	3.4	2.5	1.7
	0.2	2.2	1.7	1.1
10	0.4	3.0	2.2	1.5
	0.3	2.2	1.7	1.1
	0.2	1.5	1.1	0.8

*Crop energy output/net crop production task energy input.

crops and the home garden, from small domestic livestock, and from wild food-gathering and hunting, all of which are activities that do not normally appear in the crop production energy budget.

■ 10.5 THE ENERGETICS OF DRAFT ANIMALS

■ 10.5.1 *The energy cost of maintaining draft animals.* For detailed information on the energetics of draft animals, in one environmental situation at least, we are indebted to Odend'hal (1972), who studied 3,770 cattle in an area of India of monsoon climate with an average rainfall of 1,250 to 1,750 mm. The area is intensively utilized for cropping and represents the extreme situation (but typical for India) in which, in the absence of substantial areas of grazing land, draft cattle or buffalo are maintained largely on crop products and crop residues. Of the average daily energy intake of the cattle studied, 66 percent came from sugarcane tops, 14 percent from rice straw, 14 percent from rice hulls, and 6 percent from oilseed cake. (It is probable that the area was partially irrigated, and obviously some perennial crops were grown, but there is no reason to suppose that the general principles that emerge would not apply equally to intensive rainfed annual cropping.)

Odend'hal estimated the average daily intake of draft cattle as:

Adult males (30.5 percent of population)	91 MJ day^{-1}
Adult females (33.6 percent of population)	57 MJ day^{-1}
Immature cattle (35.9 percent of population)	38 MJ day^{-1}

In a study of the energetics of Hong Kong agriculture, Newcombe (1976) used a figure of 84 MJ day^{-1} for working cattle.

Nearly all the adult males were working animals. Thus on average, for each adult male, an additional 1.1 adult female and 1.2 immature cattle were maintained. Hence the total daily energy cost of maintaining continuously one adult working male, i.e., providing for its replacement, was:

1 adult male at 91 MJ day^{-1}	91 MJ day^{-1}
1.1 adult females at 57 MJ day^{-1}	63 MJ day^{-1}
1.2 immature cattle at 38 MJ day^{-1}	46 MJ day^{-1}
Total	200 MJ day^{-1}

If we assume an average farm family of six persons, equivalent to four adult males with, say, a daily energy consumption of 10 MJ each, then total daily human energy consumption is 40 MJ. If the farm unit maintained a pair of working cattle and an average proportion of followers in the form of breeding females and young stock, total daily energy consumption by stock would be 400 MJ, ten times that of the farm family. This simple estimate illustrates the intense effort that has to go into the maintenance of draft animals in areas where only a very small proportion of their feed requirements can be obtained from "free" grazing. It should be borne in mind, however, that most of the material fed cannot be consumed by man, and in this case the input is merely the human energy expended in harvesting and feeding. However, as Boserup

(1965) points out, this can amount to almost half the total energy expended in crop production.

At the other end of the scale, in regions of low human and draft animal population density where the proportion of land carrying natural forage is correspondingly high, whether within the farm unit or on common land, draft animals and their followers may obtain all or nearly all of their energy requirement from grazing. This may demand only the minimum of human energy input; perhaps only a child with a stick to take the animals out to graze and bring them back again.

There is, naturally, a complete range of intermediate situations. Crop residues may be saved and fed in periods when natural forage is scarce or of low quality; the farm family may go out on foot, on bicycles, or with handcarts to gather natural forage from outlying areas; grain may be fed in periods of heavy work demand, etc. In many instances the farmer will not, or cannot, raise his own replacements and will maintain only working males, purchasing replacements when necessary. It is clear that the energy cost to the farm unit of maintaining draft animals is highly dependent on local factors.

■ 10.5.2 *The work output of draft animals.* The average number of days worked by draft animals in India has been estimated at 180 to 300 (Joshi and Phillips, 1953). The fairly low figure at the smaller end of the range is in many instances related to the seasonality of farm work in a monsoon climate. The rainfall pattern, with its substantial dry season, also governs the seasonal pattern of natural forage quantity and quality. In areas where this constitutes a major proportion of total energy intake, the second half of the dry season is a period of scarcity and the condition of draft animals deteriorates unless they are given supplementary feed. Fortunately it is also the period of minimum work demand.

In Odend'hal's study the average working day was six hours. The stress on draft animals in tropical climates must be emphasized. It is normal for animals to be worked only in the forenoon and perhaps also in late afternoon. This is particularly important for buffalo, which readily suffer heat stress and benefit from a midday wallow in cool water or mud. In fact, four to five hours is more normal for heavy field work. Draft animals can be used for more extended periods in lighter tasks such as pulling carts.

■ 10.5.3 *Comparison of hoe and draft animal cultivation.*

From the assumptions made in section 10.1.2, the net work energy expenditure of a pair of draft animals and a man is estimated at 13 times that of a man alone. (Draft animals are not always used in pairs; in the Philippines the large carabao buffalo is used singly.) Boserup (1965) presents input data in man-days for Indian rainfed cropping systems with draft animals, partitioning inputs into the several production tasks. With some assumptions, we can gain from these data an esti-

TABLE 10.6
Substitution of draft animal for human energy input in intensive rainfed cropping systems. India

State	Crop	(a) Net human energy input for land preparation ($MJ\ ha^{-1}$)	(b) Net human energy input (all tasks) ($MJ\ ha^{-1}$)	(c) Net animal energy input for land preparation ($MJ\ ha^{-1}$)	(d) Total net energy input [b+c] ($MJ\ ha^{-1}$)	(e) Ratio of (d) to (b)
Bombay	Wheat	78	234	585	819	3.5
Madhya Pradesh	Wheat	126	282	945	1,227	4.4
Punjab	Wheat	66	198	495	693	3.5
Bombay	Sorghum	42	150	315	465	3.1
Madras	Sorghum	36	234	270	504	2.2
Bombay	Gram	42	198	315	513	2.6
Mean		65	216	488	704	3.2 (unweighted)

SOURCE: Adapted from Boserup (1965); for method see text.

mate of the extent of substitution of animal for human energy. This has been done in Table 10.6.

The assumptions are:

a. Draft animals used only for land preparation.
b. Men worked 8 hr day^{-1}, animals 5 hr day^{-1}.
c. A pair of animals used.
d. Man-days and animal-pair-days for land preparation equal.
e. Net human work energy expenditure 0.75 MJ hr^{-1}; net animal work energy expenditure 4.5 MJ hr^{-1}.

Table 10.6 suggests that the ratio of human plus animal net energy input to human net energy input alone, for all crop production tasks, is about 3:1.

Since this conservation of human energy in land preparation occurs at what is normally the busiest time of the year for the farmer, the potential crop area per man is substantially greater when draft animals are used for ploughing than when the land is hand-cultivated. Ruthenberg (1971) presents data from comparable areas in Tanzania where the arable area per worker is 0.7 ha under hoe cultivation and 1.3 ha under draft animal cultivation. The saving of human energy in land preparation with the ox-plough may not be proportionately reflected in increased area cultivated per man, since it can have the effect of shifting the limiting input factor from primary cultivation to weeding or harvesting. Alternatively, the limiting factor may be merely that of available cropland; in which case draft animals are not being used as a human energy substitute at maximum efficiency.

■ 10.5.4 *Other energy outputs of draft animals.* Odend'hal (1972) found that of the total energetic output of the animal population in the area he studied, both draft animals and their followers, 84 percent was in dung, 12 percent in useful work (that is, power output) and 4 percent in milk, the proportion of which not consumed by calves would be available for farm family consumption. Work output was calculated assuming 200 six-hour working days per year.

The energy value of dried cattle dung is approximately 9 MJ kg^{-1} (National Council of Applied Economic Research, 1965), about half that of firewood (Reed and Lerner, 1974). For adult male cattle, Odend'hal estimated the mean daily energy output in dung as 17 MJ, equivalent to slightly less than 2 kg dry weight. Revelle (1976) quotes a range of 1.4 to 3.5 kg day^{-1}. Dung energy output for the other classes of cattle was also calculated. If it is assumed that the farm unit maintains two working male cattle and the appropriate replacement herd, total dung energy output amounts to 78 MJ day^{-1}, equivalent to over 4 kg of firewood.

The importance of dung as a source of fuel for the hard-pressed rural population of India is obvious. With lower population density and increased availability of timber, or greater affluence and hence ability to purchase alternate fuels, dependence on dung as a source of energy declines. The significance of the loss of nutrients which could be returned to cropland will be considered in the next chapter, but it may be pointed out here that in many farming systems where dung is not used as fuel little attempt is made to return it to cropland as fertilizer. This is waste, whereas in India dung substitutes for other energy sources in maintaining the domestic rural economy. Revelle (1976) has estimated that for the rural population of India as a whole the use of dung as fuel saves annually some 25 million tons of coal.

REFERENCES

Black, J. N. 1971. Energy relations in crop production—a preliminary survey. *Annals of Applied Biology* 67:272.

Boserup, E. 1965. *The conditions of agricultural growth*. Chicago: Aldine.

Brody, S. 1968. *Bioenergetics and growth*. 2d ed. New York: Hafner.

Chandra, S., Evenson, J. P., and de Boer, A. J. 1976. Incorporating energetic measures in an analysis of crop production practices in Sigatoka Valley, Fiji. *Agricultural Systems* 1:301.

Clark, C., and Haswell, M. 1970. *The economics of subsistence agriculture*. 4th ed. London: Macmillan.

Haswell, M. R. 1973. *Tropical farming economics*. London: Longman.

Joshi, N. R., and Phillips, R. W. 1953. *Zebu cattle of India and Pakistan*. FAO. Rome.

National Council of Applied Economic Research. 1965. *Domestic fuels in rural India*. New Delhi: NCAER.

Newcombe, K. 1976. Energy use in the Hong Kong food system. *Agro-Ecosystems* 2:253.

Odend'hal, S. 1972. Energetics of Indian cattle in their environment. *Human Ecology* 1:3.

Passmore, R., and Durnin, J. V. G. A. 1955. Human energy expenditure. *Physiological Reviews* 35:801.

Philipps, P. G. 1954. The metabolic cost of common West African agricultural activities. *Journal of Tropical Medicine and Hygiene* 57:12.

Reed, T. B., and Lerner, R. M. 1974. In: Abelson, P H., ed. *Energy: Use, conservation and supply*. Washington, D. C.: American Association for the Advancement of Science.

Revelle, R. 1976. Energy use in rural India. *Science* 192:969.

Ruthenberg, H. 1971. *Farming systems in the tropics*. Oxford: Clarendon Press.

Chapter 11

Semi-Intensive and Intensive Rainfed Cropping Systems: Biogeochemical Aspects

■ 11.1 The transition from shifting cultivation to more intensive cropping

■ 11.1.1 *Population density and intensification.* In section 7.7, the long-term effects of shifting cultivation on the productivity of cropland and on the ecological character of the general area being utilized for cropping were discussed. With increasing human pressure on land suitable for cropping, cultivation frequency increases through an extension of the cropping phase and a curtailment of the fallow phase. Even on land that was formerly dense forest, shortening of the fallow phase is associated with the development of savannah or scrub vegetation, since the recovery period before the next cropping phase is inadequate for the reestablishment of forest trees. Tree density is also reduced by the increasing demand for wood as fuel; furthermore, in areas extensively and fairly uniformly cropped, seed sources for the reestablishment of trees in the form of surviving pockets of forest may be greatly reduced or eliminated. Hence the shorter fallows associated with semi-intensive cultivation are normally dominated by grasses or scrub and are termed "grass fallow" or "bush fallow." The approach toward a more regular sequence of crop and fallow with intensification has given rise to the term "rotational bush fallow."

A distinction was drawn in section 7.7 between negative and positive responses to increasing human pressure on land: the negative responses, shortening of the fallow phase, increasingly drastic cultivation, changes to more exhaustive crops, merely aggravate the situation; the positive responses, manuring, soil conservation measures, the use of mineral fertilizer, are restorative in nature. The tragedy of the intensification process with increasing population density under rainfed conditions is that restorative measures are not taken by the agricultural population in general until a late stage of land deterioration. As Protheroe (1962) puts it in relation to West Africa:

> The evidence indicates that farmers in Northern Nigeria and elsewhere in tropical Africa will intensify their methods of cultivation and adopt measures of conservation only when they are forced to do so by extreme population pressure on land. These improved methods are in use . . . where there are population densities which range from 200 to 400 per square km. The real dangers of land deterioration seem to be greatest when the population densities are between 80 to 100 per sq. km. These figures indicate a critical transition stage between lower densities which allow cultivation under a system of land rotation and higher densities which demand the permanent cultivation of a large proportion of the cultivable land.

Protheroe thus suggests that the critical point, in respect of declining soil fertility and increasing erosion, comes at the transition not between shifting and semi-intensive cultivation, as we have defined them, but between semi-intensive and intensive cultivation. At this stage restorative measures, at least in the West African environment, become urgent, and the difference in consequences between implementation and neglect is at its greatest. Ruthenberg (1971) quotes an excellent example from Senegal of contrasting semi-intensive systems of growing peanuts and millet, the one exploitative and creating a highly unstable situation, the other resource-conserving (largely, it appears, from systematic folding of cattle on arable land).

■ 11.1.2 *Characteristics of the short fallow.* The vegetation that reestablishes itself on cropland that is temporarily retired from cultivation in semi-intensive systems is likely to be dominated at first by annual grasses and weeds, and later by perennial grasses and perhaps scrub. Although the capacity of grass root systems to bring subsoil nutrients to the surface appears to be much less than that of forest (see chapter 7), it is likely that this cycling will be of significance provided the fallow is long enough for deep-rooted perennials to establish and the root systems are not weakened by heavy grazing during the growing season. Because the grasses are dormant in the dry season, and apparently translocate nutrients to storage sites at the end of their wet-season growth (Norman, 1963), it is probable that if the grass fallow is burned or grazed only in the dry season its capacity for nutrient restoration is not greatly impaired in comparison with ungrazed or unburned fallow. If the fallow is grazed in a controlled fashion, with stock penned on the field and the residues returned *in situ*, then nutrient uptake from the deeper soil layers may be reduced; but at least those nutrients stored in aboveground material are largely returned to the soil, the major loss being nitrogen volatilized as ammonia from urine deposits. The presence or otherwise of dung-eating insects has a marked effect on the rate of nutrient recycling (Bornemissza, 1976).

The proportion of legumes in tropical short fallows is low, and substantial increments of nitrogen from this source are unlikely. The significance of nonsymbiotic fixation or fixation associated with grasses in the field situation has yet to be determined. Two aspects of fallow nitrogen cycling appear to be clear, however. First, leaching is negligible once a complete grass cover is established. Second, if available nitrogen is low when the cropfield is retired from cultivation, there will be an accumulation of organic carbon in grass roots and stubble during the life of the fallow which, when it is recultivated, may immobilize nitrate-nitrogen over the period in which it is being decomposed. This period does not normally extend beyond one cropping season.

In the chemical sense, therefore, the short grass fallow may be regarded as weakly restorative. In respect of its effect on soil physical condition, it has already been mentioned that the beneficial effects of a grass fallow on infiltration rates and water-holding capacity as a result of the accumulation of organic matter are transient and rarely extend beyond the first year of cropping (Wilkinson, 1975).

■ 11.1.3 *The long-term nutrient cycle.* As the fallow phase in cropping systems becomes shorter, to the stage where cropland is rested only occasionally or is cultivated every year, the significance of the fallow in the general nutrient cycle of the system declines to the vanishing point. The biogeochemical stability of the system comes to depend far more upon the balance of nutrient input and output with cropping, on the rates of accumulation and decomposition of organic matter and of solubilization and fixation of mineral nutrients, and on the degree of soil erosion. In the absence of inputs of animal manure or mineral fertilizers, such stability is achieved only at a low level of crop output, as in many parts of India. The decline in crop yield with continuous cropping, illustrated largely by examples from Africa, is considered in the following section. It should be borne in mind that the general biogeochemical situation under which most of the African experiments have been conducted is that of soils not previously subjected to intensive cropping for a long historical period. In other words, they represent a condition of biogeochemical instability.

■ 11.2 Decline in crop yield with intensive cropping

■ 11.2.1 In Africa a large number of surveys and formal experiments has been carried out on the decline in yield with continuous annual cropping. Some of the more significant have been summarized by Nye and Greenland (1960). Most of the data relating to high-rainfall forest zones have come from surveys of farmers' fields; they have the merit of direct relevance, but the relative importance of declining soil fertility and weed incidence cannot be assessed. In general, there seems little doubt that in the absence of manure or fertilizer, crop yield declines rapidly with successive years of cultivation in tropical regions with rainfall in excess of 1,500 mm and on forest Oxisols and Ultisols. Figure 7.2 shows some examples.

On the other hand, in savannah climates much of the information is from long-term formal experiments, the results of which are less immediately applicable. The experiments have had the advantage, however, of soils kept weed-free, so that their results can generally be ascribed to soil chemical or physical factors. Broadly speaking, yields decline more slowly with continuous cropping on the more fertile soils associated with savannah climates. A summary of the results from a long-term experiment at Kano, Northern Nigeria (rainfall 900 mm), in which peanuts, millet, and sorghum were grown continuously for 25 years, is shown in Table 11.1 (Dennison, 1961; Nye and Greenland, 1960).

In a series of 2^5 fertilizer trials in Ghana in a range of rainfall zones, plots were planted to a range of crops (Djokoto and Stephens, 1961a). In general, the major limiting nutrients with successive cropping at high-rainfall locations were phosphorus followed by potassium, and at lower-rainfall locations nitrogen followed by phosphorus. Comparing crops, the main deficiencies in cereals were nitrogen and phosphorus, in cassava potassium, and in peanuts phosphate and lime.

A striking example of the long-term effects of short-fallow cropping systems on soil characteristics is shown in Table 11.2. The data are from the Casamance area of South Senegal in a 1,300 mm rainfall region. The progressive decline in organic carbon, organic nitrogen, carbon/nitrogen ratio, available soil water and exchangeable nutrients over a period of a century from the time the land was first cultivated is very clear indeed. Table 11.2 is an excellent illustration of a region of semi-intensive cropping systems that has not yet achieved biogeochemical stability.

The above data are merely illustrations from the very large number of surveys and experiments that have been carried out. There have been other equally reliable studies that have shown a less dramatic or negligible decline in yield with continuous cultivation: for example, with

TABLE 11.1
Declining yields of continuous cropping. Nigeria

5-year cropping period	Peanuts (kernels kg ha^{-1})	Millet (grain kg ha^{-1})	Sorghum (grain kg ha^{-1})
1931–35	1,015	920	540
1936–40	785	455	330
1941–45	700	320	105
1946–50	320	545	90
1951–55	510	330	Discontinued

(In the above experiment, manure was applied to other plots; its effect is discussed later.)

TABLE 11.2
Deterioration in soil chemical and physical conditions with time under semi-intensive cropping. Senegal

Duration of cropping period from original forest clearing (yrs)	Organic matter			Mean available soil water (percent)	Cations (meq 100g soil^{-1})		
	Carbon (percent)	Nitrogen (percent)	C/N ratio		Ca	Mg	K
0	16.5	0.90	18.3	4.1	5.0	1.7	0.07
3	13.8	0.79	17.5	4.7	2.7	1.2	0.07
12	11.6	0.68	17.0	3.7	2.2	1.0	0.04
46	6.8	0.43	15.8	2.8	1.4	0.5	0.04
90	5.0	0.35	14.3	3.3	1.0	0.5	0.04

SOURCE: Mouttapa (1974); after Suban, no reference.

cotton in Uganda over twenty years in 1,250 mm rainfall (Nye and Greenland, 1960) and with cotton and peanuts in North Australia over ten years in 900 mm rainfall (Norman, 1966). In other experiments, nutrients other than nitrogen and phosphorus appear to be limiting: for example, in a series of trials in Uganda the main deficiency with continued cropping was potassium (Stephens, 1969).

■ 11.2.2 The evidence for an effect of declining soil physical condition on yield with continued cropping is far less reliable, if only for the reasons that agronomically significant parameters of soil physical condition are not clearly defined (except perhaps available soil water) and because controls analogous to the fertilizer treatments applied when testing for nutrient deficiency are difficult to establish.

Table 11.2 shows an effect of long-continued cropping on available soil water, but short-term experiments on soil physical attributes have been less conclusive. Observers have frequently noted a general deterioration in tillage characteristics, but in the North Australian experiments, for example Norman (1966), this was not associated with declining crop yield. Pereira et al. (1967) demonstrated a fall in infiltration rate with continued cropping in Kenya.

It would be expected that soil physical condition would be closely related to organic matter content, but we must beware of regarding all fractions of soil organic matter as being equally labile (Nye, 1963). In general, it appears that on land that is cleared of natural vegetation and then cropped intensively there may be fairly rapid decomposition of the more labile fraction, but that the less readily decomposed fraction of the soil organic matter becomes in time relatively stable. Furthermore, the fluctuations in the more labile fraction occasioned by periods under crop or fallow are transient. Such a situation, of course, presupposes a condition where the rate of soil erosion is not such as to remove whole layers of soil.

■ 11.3 The role of nitrogen

■ 11.3.1 As a component of the biogeochemical cycle, nitrogen merits special attention since it is the most labile of soil nutrients and because the available nitrogen status of the soil may be readily modified by agronomic practice. Once other nutrient requirements are satisfied, crop yield is very much dependent on the level of available nitrogen and, indeed, often displays a very close positive relation to total nitrogen uptake by the crop if water is not limiting.

■ 11.3.2 *The pattern of nitrogen transformation and movement.* In a savannah climate, the mineralization of soil organic nitrogen is arrested in the dry season because of the dryness of the soil. By the time the rains begin, temperatures are high and as the profile is recharged rapid mineralization begins. It appears that rate of mineralization is increased when the soil has been previously subjected to intense drying (Birch, 1958). Furthermore, land preparation, resulting in greater aeration and aggregate disruption (Rovira and Greacen, 1957) will hasten it. In a free-draining soil, continued rain and drainage through the profile will leach some of the flush of mineral nitrogen, and crop uptake will to some extent depend on time of planting and the rate of root extension to deep soil layers. In heavy soils, leaching will be less important but the possibility of loss through denitrification under anaerobic conditions is increased. The major exception to the above, on all soil types, is when large amounts of carbonaceous material from fallows or crop residues are incorporated, resulting in nitrogen immobilization.

The depth to which nitrate will be leached will be dependent on the permeability of the soil and the amount of rainfall (Wetselaar, 1962) or, more accurately, on rainfall less evaporation, that is, percolating rain (Jones, 1975). In savannah climates, high yields are normally associated with early planting, and for nonlegumes this relation is often associated with reduced nitrogen uptake by late-planted crops because of early leaching, though this is disputed (Macdonald, 1968). However, fresh bursts of mineralization may occur later in the season, which may prevent a decline in yield with later planting (Jones, 1975).

In general, however, the greater part of the total nitrogen mineralized over the cropping period is made available in the first 2 to 3 months of the wet season, and except when the available nitrogen concentration is extremely high through the addition of fertilizer, crop uptake toward maturity is low. In grain crops maximum aboveground nitrogen yield is normally reached at about anthesis and may decline thereafter. It is therefore not surprising that in many relay crop systems the late-planted crop is a legume, since it has to establish itself in a situation of strong competition for available nitrogen.

Far less is known about the mineralization pattern of soil nitrogen in the wet tropics. It is probable that the process is intermittently continuous, slowing down during dry spells and increasing when they are broken by further rain. Liability to leaching losses in free-draining soils and to denitrification losses in poorly drained soils will be greater owing to higher rainfall. In the absence of the extreme drying of the savannah winter, it is probable that the flush of mineralization at the start of a wet period will be less marked. On the other hand, the total period of the year when the crop root zone is adequately moist for mineralization is longer than in savannah climates; hence if other chemical and physical conditions are comparable, the amount of nitrogen mineralized in the course of the year is likely to be greater.

■ 11.3.3 *Nitrogen input and output by legumes.* It has been indicated in section 4.2.3 that tropical crop legumes do not necessarily make any significant addition to the total nitrogen store in the crop root zone. However, there have been a surprisingly small number of adequately controlled studies on the nitrogen balance of tropical cropping sequences with annual legumes. One of the few is that of Wetselaar (1967) in North Australia, on a free-draining soil in a savannah climate with a rainfall of 900 mm. The example is particularly useful in demonstrating to what degree the removal or return of crop material can influence soil nitrogen status.

Wetselaar grew four annual legumes each for three successive years and compared them with a three-year bare fallow. The crops were:

a. Townsville stylo (*Stylosanthes humilis*). A self-seeding annual forage legume; cut for hay and forage removed.
b. Peanuts. Nuts only removed; residues left.
c. Guar (*Cyanopsis tetragonoloba*). A grain legume; grain only removed and residues left.
d. Cowpeas. Cut for hay and forage removed.

Figures 11.1 and 11.2 summarize the results. Figure 11.1 shows over three years:

a. Nearly 400 kg ha^{-1} N removed in cowpea; none returned.
b. About 200 kg ha^{-1} N removed in T. stylo; a little returned in aftermath.
c. About 150 kg ha^{-1} N removed in peanuts; about 50 kg ha^{-1} N returned.
d. About 50 kg ha^{-1} N removed in guar; nearly 200 kg ha^{-1} N returned.

Figure 11.2 shows the nitrogen balance sheet after three years. Only with guar and T. stylo was the original topsoil nitrogen status—organic plus mineral nitrogen—maintained. Three successive cowpea or peanut crops, with the bulk of their plant nitrogen removed from the field, resulted in a significant fall in organic plus inorganic nitrogen in the topsoil.

It is hence stressed once more that statements about the nitrogen-restoring power of legumes in tropical crop sequences must be treated

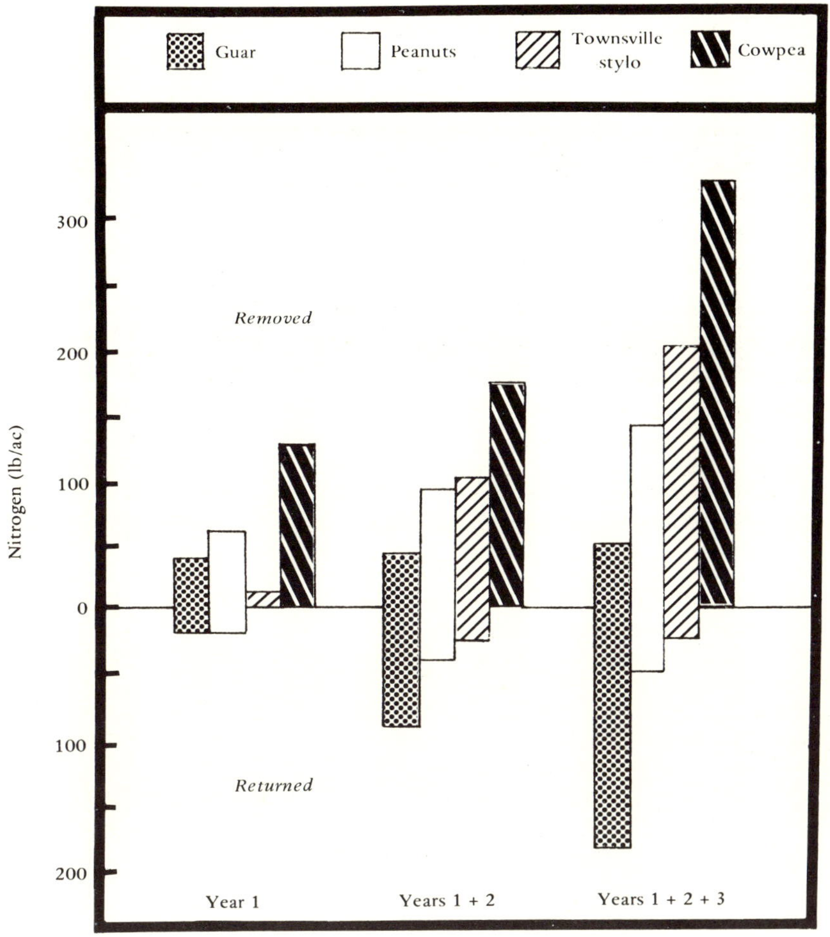

Figure 11.1. Crop nitrogen yield of four annual legumes over three years. N. Australia (Wetselaar, 1967).

with caution and qualified by identifying the crop that is being grown and the manner in which its product components are handled.

■ 11.4 ANIMAL MANURE

■ 11.4.1 Animal manure, or "kraal" manure as it is known throughout Africa, is used in semi-intensive and intensive rainfed cropping systems with varying degrees of purposefulness. At the minimum, the accumulation of dung from domestic animals housed near the domicile will be used on the home garden for fruit trees and vegetables, etc. Perhaps the next stage in intensification is its application to cropfields near the domicile, which may result in the concentric pattern of fertility

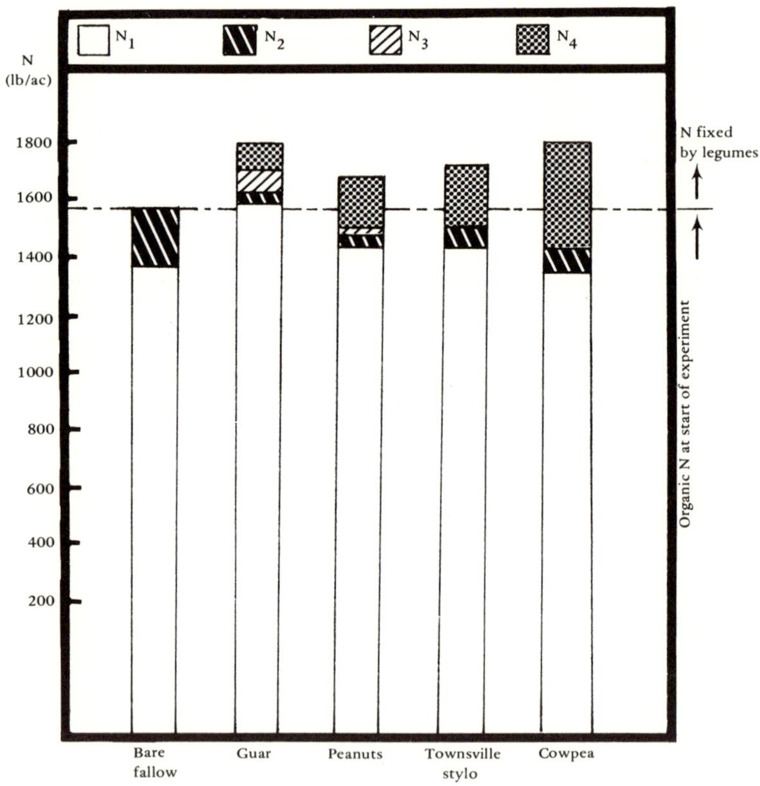

Figure 11.2. Nitrogen balance after three years of cropping with four annual legumes. N_1—topsoil organic N; N_2—NO_3-N in soil profile; N_3—plant N returned in third season; N_4—plant N removed. N. Australia (Wetselaar, 1967).

and consequent cultivation frequency already referred to. Mention has already been made of the practice in American mixed crop and livestock systems of siting crop areas downslope from cattle yards to benefit from nutrient-rich runoff water.

Further developments are likely to be the planned application of manure to those cropfields in greatest need, or at a particular point in a crop sequence, and the penning of livestock on cropfields to consume residues. In West Africa, mutual arrangements are made between pastoralists and cultivators for the former to pen their stock on the cropfields of the latter: the pastoralist gains feed in the form of crop residues while the cultivator benefits from the dung and urine. In the intensive rainfed cropping regions in North Nigeria crop farmers may purchase kraal manure. The ultimate development is seen in the Indian subcontinent where dried cattle and buffalo dung is used as fuel rather than as fertilizer. Apart from manure, compost may be made from grass, green crop residues, household waste, etc., in some highly intensive cropping systems in Asia and Africa.

■ 11.4.2 The nutrient content of dung depends greatly on the nutrient content of the animal's feed intake; standard textbook figures are therefore of little value since they are based on well-fed or overfed (by tropical standards) production animals. Phillips (1956) gives ranges of actual sample values for kraal manure and compost in West Africa (see Table 11.3).

TABLE 11.3
Nutrient content of kraal manure and compost

	Kraal manure % dry weight	Compost % dry weight
Nitrogen	1.03–2.12	0.43–0.91
Phosphorus	0.11–0.28	0.04–0.12
Potassium	0.97–2.30	0.17–1.12

■ 11.4.3 There have been many studies in Africa of the effect of manure on crop yield under semi-intensive and intensive cropping. As an example, in the long-term continuous cropping experiment in N. Nigeria referred to earlier (section 11.2.1), peanuts, millet, and sorghum were grown with and without 6.5 t ha^{-1} of manure. The results are shown in Figure 11.3. Manure maintained the yield of millet over eighteen years, and almost maintained the yield of sorghum; peanuts showed a response to manure but the average rate of yield decline was similar to that without manure, perhaps the result of inadequate phosphorus or calcium. Djokoto and Stephens (1961b) found that manure applied at 5 to 10 t ha^{-1} in long-term fertilizer trials in Ghana under continuous

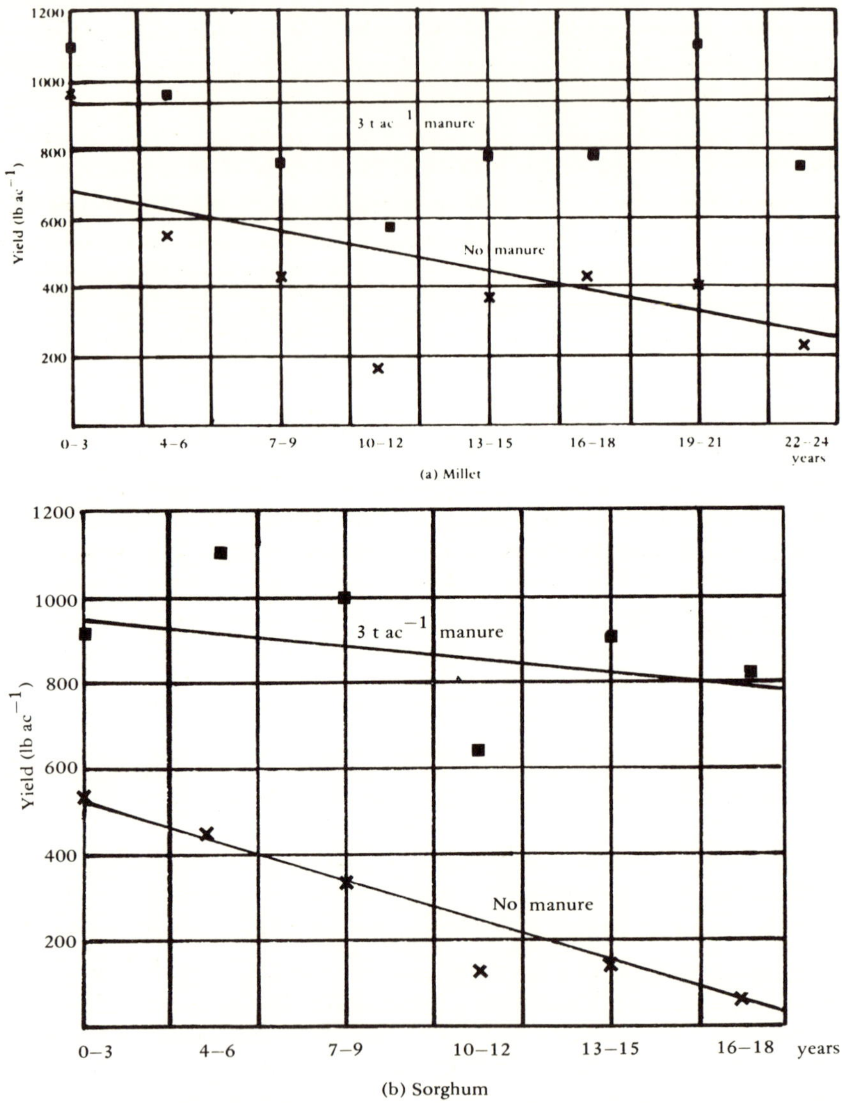

Figure 11.3a,b. Changes in crop yield under continuous cropping with and without manure (Dennison, 1961).

cropping increased organic carbon and nitrogen, available phosphorus, and exchangeable potassium and magnesium.

It is generally agreed that the value of manure lies largely in its nutrient content and that its effects on soil physical structure are limited and transient in the tropics, though if applied annually some physical benefit is sustained. It may also have a beneficial effect in reducing soil acidity (Heathcote, 1969). Djokoto and Stephens (1961a), commenting upon the superiority of manures to inorganic fertilizer in their long-term trials in Ghana, particularly apparent after some years of cropping, suggested that "a residue of unexplained benefit remains which is probably connected with the balanced nutrient reserves slowly released during the crops' growth by the decomposition of organic matter." In

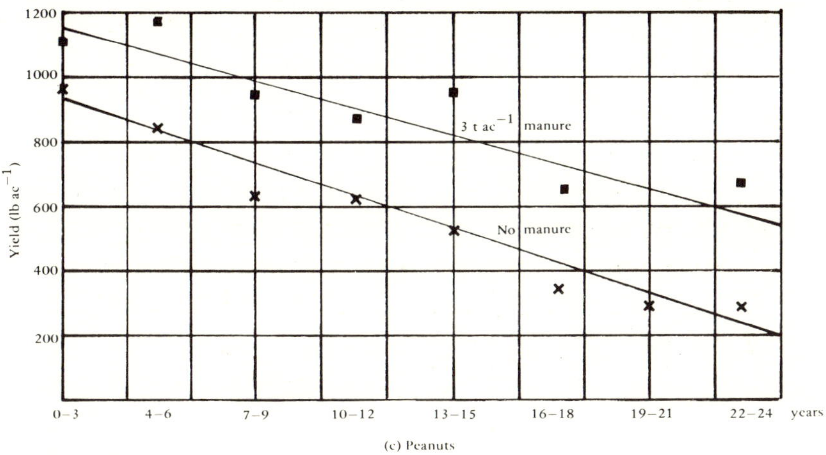

Figure 11.3c. Changes in crop yield under continuous cropping with and without manure (Dennison, 1961).

other words, a nutrient response but with the added benefit of slow release.

■ 11.4.4 *Nutrient losses from using dung as fuel.* Revelle (1976) quotes figures of dung output per head of cattle in India of 1.4 to 3.5 kg day^{-1} (dry), of which an estimated 22 to 75 percent is used as fuel. An average figure of two-thirds used as fuel has been estimated by the National Council of Applied Economic Research (1965). If we apply the values of Phillips (1956) for nutrient content, the amounts of nutrient produced in dung per head per year are 5.3 to 27.1 kg nitrogen, 0.6 to 3.6 kg phosphorus, and 5.0 to 29.4 kg potassium. For the average farm "herd" where two adult males are used for draft in the Indian survey by Odend'hal (1972), that is 2 adult males, 2.2 adult females and 2.4 immature cattle, total nutrient output in dung, using Odend'hal's figures for relative dung output by the different classes of cattle, would amount to:

Nitrogen	24–125 kg $farm^{-1}$ yr^{-1}
Phosphorus	2.8–16.6 kg $farm^{-1}$ yr^{-1}
Potassium	23–136 kg $farm^{-1}$ yr^{-1}

Applying the estimates given above of the proportion used as fuel (22 to 75 percent), the extreme ranges of nutrients not directly returned as dung but burned become:

Nitrogen	5–94 kg $farm^{-1}$ yr^{-1}
Phosphorus	0.6–12.4 kg $farm^{-1}$ yr^{-1}
Potassium	5–102 kg $farm^{-1}$ yr^{-1}

It is clear that in the upper sector of these ranges the nutrient drain through using dung as fuel is significant. However, only the nitrogen (and sulfur) components represent a loss to the farm system as a whole, since presumably the ash from domestic fires is used as fertilizer, at least on the home garden, and hence phosphorus, potassium, and other ash minerals are recycled.

REFERENCES

Birch, H. F. 1958. The effect of soil drying on humus decomposition and nitrogen availability. *Plant and Soil* 10:9.

Bornemissza, G. E. 1976. *The Australian dung-beetle project.* Australian Meat Research Committee Review no. 30.

Dennison, E. B. 1961. The value of farmyard manure in maintaining fertility in northern Nigeria. *Empire Journal of Experimental Agriculture* 29:330.

Djokoto, R. K., and Stephens, D. 1961a. Thirty long-term fertilizer experiments under continuous cropping in Ghana. I. Crop yields and responses to fertilizers and manures. *Empire Journal of Experimental Agriculture* 29:181.

Djokoto, R. K., and Stephens, D. 1961b. Thirty long-term fertilizer experiments under continuous cropping in Ghana. II. Soil studies in relation to the effects of fertilizers and manures on crop yields. *Empire Journal of Experimental Agriculture* 29:245.

Heathcote, R. G. 1969. Soil fertility under continuous cultivation in northern Nigeria. I. The role of organic manures. *Experimental Agriculture* 6:229.

Jones, M. J. 1975. Leaching of nitrate under maize at Samaru, Nigeria. *Tropical Agriculture (Trinidad)* 52:1.

Macdonald, A. S. 1968. The early planting controversy. *World Crops* 20:78.

Mouttapa, F. 1974. Soil aspects in the practice of shifting cultivation in Africa and the need for a common approach to soil and land resources evaluation. In: *Shifting cultivation and soil conservation in Africa.* FAO Soils Bulletin no. 24. Rome.

National Council of Applied Economic Research. 1965. *Domestic fuels in rural India.* New Delhi: NCAER.

Norman, M. J. T. 1963. The pattern of dry matter and nutrient content changes in native pastures at Katherine, N.T. *Australian Journal of Experimental Agriculture and Animal Husbandry* 3:119.

Norman, M. J. T. 1966. *Katherine Research Station 1956–64: A review of published work.* Commonwealth Scientific and Industrial Research Organization, Australia, Division of Land Research Technical Paper no. 28.

Nye, P. H. 1963. Soil analysis and the assessment of fertility in tropical soils. *Journal of Science in Food and Agriculture* 14:277.

Nye, P. H., and Greenland, D. J. 1960. *The soil under shifting cultivation.* Harpenden, England: Commonwealth Bureau of Soils. Technical Communication no. 51.

Odend'hal, S. 1972. Energetics of Indian cattle in their environment. *Human Ecology* 1:3.

Pereira, H. C., Hosegood, P. H., and Dagg, M. 1967. Effects of tied ridges, terraces and grass leys on a lateritic soil in Kenya. *Experimental Agriculture* 3:89.

Phillips, T. A. 1956. *An agricultural notebook.* London: Longman.

Protheroe, R. M. 1962. Some observations on desiccation in North-western Nigeria. *Erdkunde* 16:112.

Revelle, R. 1976. Energy use in rural India. *Science* 192:969.

Rovira, A. D., and Graecen, E. L. 1957. The effect of aggregate disruption on the activity of micro-organisms in soil. *Australian Journal of Agricultural Research* 8:659.

Ruthenberg, H. 1971. *Farming systems in the tropics.* Oxford: Clarendon Press.

Stephens, D. 1969. Changes in yields and fertilizer responses with continuous cropping in Uganda. *Experimental Agriculture* 5:263.

Wetselaar, R. 1962. Nitrate distribution in tropical soils. III. Downward movement and accumulation of nitrate in the subsoil. *Plant and Soil* 16:19.

Wetselaar, R. 1967. Estimation of nitrogen fixation by four legumes in a dry monsoon area of north-western Australia. *Australian Journal of Experimental Agriculture and Animal Husbandry* 7:518.

Wilkinson, G. E. 1975. Effect of grass fallow rotations on the infiltration of water into a savannah zone soil of northern Nigeria. *Tropical Agriculture (Trinidad)* 52:97.

Chapter 12

Semi-Intensive and Intensive Rainfed Cropping Systems: Hydrological Aspects

■ 12.1 Soil erosion

■ 12.1.1 With increasing cultivation frequency, the potential threat of runoff-induced soil erosion increases. The main factor involved is the increase in the proportion of years when the soil surface is in the exposed arable condition rather than in the fallow condition with a closed vegetation cover. In addition, increasing intensity of cultivation may in some instances bring about a reduction in the infiltration capacity of the soil surface. As an example of the latter effect, Pereira et al. (1967) in Kenya measured the progressive decline in infiltration for a standard period and intensity of rainfall with increasing years of continuous cropping:

after 1 year of cropping	94 percent infiltration
after 2 years of cropping	89 percent infiltration
after 3 years of cropping	91 percent infiltration
after 4 years of cropping	75 percent infiltration

In a savannah climate, vegetative cover on a cropfield is slight or negligible during the early wet season, and will not approximate to the cover of natural perennial vegetation until mid–wet season. Even then, owing to low plant density, complete rainfall interception may not be achieved. The lack of cover at the start of the wet season is particularly important since in many savannah climates the erosivity of early rains from violent convectional storms is greater than that of the main rain periods, which arise from weak depressions and troughs of low pressure. In more equatorial climates, where a more extended cropping season is possible, the proportion of the year in which the cropfield has an inadequate vegetative cover is lower than in savannah climates; on the other hand, rainfall is heavier and the rain period longer.

On soils with sandy surfaces, infiltration rate is high, but against this soils of such texture are often highly mobile under the impact of rain. On nonswelling clay soils, infiltration rate at the beginning of the rains, before ploughing, is often very low because the surface has been baked hard during the dry season. Cracking clay soils accept water readily at the start of the season into deep cracks, but once the profile is recharged and the cracks sealed, infiltration drops sharply and runoff and erosion may be heavy.

■ 12.1.2 Many quantitative small-scale studies have been made of runoff and soil loss from tropical cropfields or small cropped catchments of differing soil type, crop, and agronomic management. Some examples were given in chapter 2 (Tables 2.2–2.4). Nye and Greenland (1960) have summarized others, and Table 7.7 is one instance of the contrast between fallow land, cropland, and bare soil. The actual amount of runoff and soil loss recorded depends greatly on local conditions, and there is little point in presenting a mass of data. The amounts recorded are also influenced by the scale of measurement: in a large catchment monitored only at its outlet there may be runoff and soil loss that is absorbed or deposited in the catchment and not measured. For example, compare Tables 2.2 and 2.4.

As an example of larger-scale effects, Pereira and Hosegood (1962) compared the water yield and soil loss over a year from two small headwater valleys in Kenya of average slope 1 in 4. One was grass and shrub in its upper part and cropland without soil conservation measures in its lower part; the other was uncultivated, being grass and shrub in its upper part and forest in its lower part. The figures for soil loss given below are for steady flow conditions; losses immediately after storms were not measured.

	cultivated catchment	uncultivated catchment
Water yield (mm):	1,060	610
Soil loss (kg ha^{-1}):	719	174

Apart from the difference in soil loss, which would probably have been proportionately wider had storm losses been taken into account, the different capacities of forest vegetation and cropland to absorb and retain water for slow gentle drainage are noteworthy.

Potentially one of the most serious situations for soil erosion is created when intensive rainfed cultivation is undertaken on sloping land in high-rainfall regions where the valleys are occupied by a dense population practicing intensive agriculture. As the catchment is degraded, not only is less water retained on the slopes for crop growth, but sporadic flooding becomes more frequent and the irrigation and drainage reticulation of the valleys becomes choked with silt.

■ 12.2 Water management and soil conservation practices

■ 12.2.1 It must be borne in mind that water retention or runoff and soil retention or loss are two separate processes. In semi-arid tropical climates water retention and soil conservation are equally desirable goals that are often attempted with the same techniques or structures. In the high-rainfall tropics, however, the hydrological aim is frequently that of runoff disposal, which has to be achieved with the minimum of soil loss. In the summary below, based largely on Kampen (1974), the techniques of water management and soil conservation are considered together.

■ 12.2.2 *Mulching.* Although surface mulching as a means of conserving water and reducing soil loss has been found to be highly effective in regions of relatively high rainfall and short intermittent dry spells, as demonstrated by work at IITA (Lal, 1975), it appears to be less useful in more arid environments. The extensive studies by the All-India Coordinated Research Project for Dryland Agriculture have not shown any decisive advantage. Furthermore, in the more arid environments the grass and crop residues required to provide an adequate mulch cover are likely to be in demand as forage for ruminants. An example of the benefits from surface mulching at IITA, Nigeria, is given in Table 2.4. Vertical mulching, that is, the incorporation of mulch in a band in the soil, has been shown to be effective in combination with tied ridges for cotton (Lawes, 1961) but not for the more quick-maturing crops sorghum and peanuts (Lawes, 1966).

■ 12.2.3 *Contour bunds.* Broad contour bunds are widely used in India, particularly on heavy black soils, to retain water above the bund and hence to reduce runoff and erosion and to increase water supply to the crop. "Although contour bunds, if well designed and properly maintained, and if provided with wasteweir structures, are very effective as a soil conservation measure, their usefulness as a water conservation technique under the highly erratic rainfall conditions prevailing in much of the semi-arid tropics is not supported by research data" (Kampen, personal communication). They may actually reduce crop yield when it is measured on the basis of total potential crop area. The uncropped area occupied by the bund itself, the lower fertility of the borrow area immediately above from which topsoil is removed to build the bund, and the reduction in yield which may occur above the bund owing to inundation in periods of heavy rain all act to reduce overall crop yield. Infiltration is increased only on a relatively small proportion of the total cropland. A further disadvantage of the contour bund is the likelihood of localized catastrophe should it break. Graded bunds appear to be little more effective in retaining water for increased crop production than contour bunds.

■ 12.2.4 *Graded furrows, waterways and farm ponds.* For many centuries in India, runoff water has been collected from catchments of varying size and stored in reservoirs called "tanks" for domestic use or for gravity irrigation below the tank, particularly of rice. Historically, however, the surface flow of water into the tanks has not been controlled. As a result of poor maintenance, overgrazing, and lack of erosion control in the catchment, many of these are now ineffective owing to siltation and a high surface-to-volume ratio. Another long-established practice is the recovery of drainage water from wells for irrigation and domestic use. The irrigation aspects of these systems will be considered later; the point here is that while such water-harvesting patterns benefit those fortunate enough to own the irrigable land, runoff is not controlled and much-needed water is lost from the upper catchment under rainfed crops. Modern adaptations of this established practice aim at systems that can serve a single farm or small group of farms and which may benefit rainfed and irrigated cropping areas alike.

At ICRISAT, Hyderabad, considerable progress is being made in the establishment, in small cropped catchments of varying size (see Table 2.2), of systems of beds or ridges separated by furrows, graded at about 0.5 to 0.8 percent and leading into graded grassed waterways and thence to small farm ponds. The water thus stored may be used for irrigation of fields below the pond or pumped back on to the rainfed crops in the catchment for supplementary "life-saving" irrigation during dry periods (Kampen, 1974).

■ 12.2.5 *Bench terraces.* Agricultural man in the tropics throughout history has, under pressure of population increase in areas of rolling or abrupt topography, developed various systems of land terracing, with or without retaining banks, to increase infiltration and retention and to reduce runoff and erosion. In regions of high rainfall in southeast Asia, bench terraces with a retaining bank have enabled farmers to cultivate wet rice safely on very steep slopes, a practice seen at its maximum development in Indonesia and the Philippines. Such systems will be examined in the section on irrigated cropping but are mentioned here to point out that whereas the banked, terraced field is the most stable form of land geometry in high rainfall regions, it has its disadvantages for rainfed nonflooded crops. Such crops may be grown on the flat on stored soil water, but if they are grown during the rainy season it is usually necessary to establish internal surface drainage systems within the field by forming beds or ridges.

■ 12.3 Rainfed cropping patterns in relation to soil water

■ 12.3.1 *Fallowing for water conservation.* In warm temperate climates of winter rainfall dominance or with rainfall in both summer and winter, it is common practice to maintain the land in a clean-cultivated condition for a year or half-year to accumulate soil water for a subsequent crop. Thus in semi-arid Mediterranean climates a typical two-year rotation is fallow–cereal; in semi-arid climates with a better rainfall distribution land may be fallowed in winter for a summer crop or in summer for a winter crop. (In this context, "fallowing" signifies clean cultivation without cropping.) Cropping strategy for alternating crop–fallow sequences of subtropical climates has been modeled by Fitzpatrick and Nix (1969).

In high-rainfall tropical climates such tactics are, of course, unnecessary. In semi-arid savannah climates, fallowing land throughout one wet season with the aim of conserving subsoil water for a crop in the following wet season is virtually unknown, for two reasons. First, it requires the farm unit to expend cultivation energy with a return in the form of a crop only every second year; second, the long period of evaporative loss during the dry season reduces effective soil water storage to low levels. For the annual crop of limited rooting depth, the hydrological slate is virtually wiped clean each year.

On the other hand, over large areas in India on heavy Vertisols in a rainfall range of 600 to 1,500 mm, it is common practice to clean-fallow land through the wet season and to crop it early in the dry season, e.g., to sorghum or wheat, the crop being grown on stored water (Kampen, 1974; Krantz, 1974). Reasons advanced for this strategy include the difficulties of wet-season land preparation with oxen on the heavy soils, the liability of the crop to temporary flooding, and the probability of drought periods, particularly in the more arid regions and in the early phase of the monsoon. Furthermore, in the drier areas, growing a wet-season crop may leave insufficient stored water for a dry-season crop. The potential for erosion from bare soil of low infiltration rate exposed to the whole of the monsoon is high, and one can only conclude that this is a system of soil water conservation imposed by the problems of cropping in the wet season.

In the following sections, the general principles of cropping tactics over the range of tropical rainfall regions are discussed. The relations between crop patterns and length of the growing season over a transect in W. Africa are considered in Cochemé and Franquin (1967) (see chapter 2); a compendious account of cropping patterns over a whole subcontinent is to be found in a publication of the Indian Council of Agricultural Research (1970). In addition, the recent publication of the American Society of Agronomy (1976) is a mine of information.

■ 12.3.2 *Cropping tactics in short-season savannah climates.* In savannah climates with a limited growing period, owing to the combination of a short wet season and poor soil water storage, cropping tactics are simple since there is little room for manoeuvre. Monoculture, mixed cropping, or intercropping may be practiced, but the duration of the period of available water is such that all crop growth and development may have to be completed in 100 days or so. Relay cropping is scarcely possible and double cropping out of the question. The crops grown are normally cereals and legume grains since most tuber crops require a longer growing period. Adaptive tactics merely involve matching crop duration to available soil water duration; as we have seen, this has resulted in the evolution of photosensitive cultivars which, in their home latitudes, come to maturity during the period of soil water reserve utilization (Bunting and Curtis, 1968).

■ 12.3.3 *Cropping tactics in longer-season savannah climates.* An extension of the period of availability of soil water, either through an extended rainfall season or because of better soil water storage characteristics, permits more complex within-season cropping tactics. There are three basic systems, but all have a common rationale: the choice of crop on the basis of maturity type and temperature tolerance, and the choice of time of planting, to achieve the best fit between the seasonal patterns of crop water demand and soil water availability. The three types are:

a. A single long-season crop or mixture of crops.
b. Intercropping with a short and long season crop.
c. Relay cropping with short or medium-season crops.

Typical examples of type (a) are cropping with long-season cultivars of cotton in Africa and India, or with cassava in Southeast Asia and elsewhere. Cassava has within its ecotypic range cultivars more xerophytic and of more limited crop duration than most other tuber crops. A characteristic type (b) pattern in India is intercropping sorghum, a short-season crop, with pigeon pea (Krantz, 1974). The later pigeon pea is tolerant in its early stages of competition for light from the faster-growing sorghum; after the cereal is harvested late in the wet season the legume continues growing and matures on stored water in the dry season. Maximum leaf area and water demand of the crop mix is reached in late wet season at the time of maximum water availability.

Relay cropping patterns, type (c), usually combine an early summer cereal or legume relay-cropped with a similar crop. It is more common for the two crops grown to comprise one legume and one nonlegume than two of either type; for example, corn-soybeans in Indonesia. In semi-intensive cropping systems in Africa where a differentiation of microenvironment is achieved by mounding, medium-season tuber crops may be grown on the mounds and short-season cereals or legumes planted later between the mounds.

As the growing season is further extended, double cropping, i.e., without any relay overlap, becomes possible using medium- to short-season cultivars, but such tactics are more characteristic of the hydrological situation described below.

■ 12.3.4 *Cropping tactics in extended rainfall regions.* The rather vague phrase "extended rainfall" is used here as a catchall for climatic patterns that are distinguishable into three types:

a. Extended single-peak savannah rainfall climates
b. Twin-peak subequatorial rainfall climates
c. High-latitude monsoon climates with a winter rainfall component.

In all instances the total duration of the period of available soil water throughout the year is such that two successive crops can be grown. In this situation the options for cropping widen so greatly that to single out examples is not particularly useful. In Africa, especially in extended single-peak or twin-peak rainfall regions, very complex mixed cropping, intercropping, and relay cropping patterns of tropical crops are found (Okigbo and Greenland, 1976). Long-season tuber crops become more prominent; also perennial or semiperennial crops, e.g., bananas, sugarcane, which can readily survive the relatively short dry period or periods.

Climatic subtype (c) deserves particular attention since it is typical of the northern valleys of India, where, with cool winters of low evaporative demand and a winter rainfall component, two-crop rainfed systems comprising a wet-season or *kharif* crop and a winter or *rabi* crop are typical (Krantz, 1974; ICAR, 1970). The summer crop is of tropical adaptation: corn, sorghum, millet, peanuts, etc.; the winter crop of temperate adaptation: chickpea, wheat, barley, safflower, etc. On soils of good water retention, such double cropping patterns are feasible even where there is little or no winter rain; the second crop is grown wholly on stored water.

■ 12.3.5 *Cropping tactics in wet tropical climates.* Once the limitations of a recurrent and substantial dry period are removed, the potential for complex cropping becomes from the hydrological point of view almost infinite, provided drainage is good. In low-lying situations of poor drainage, however, the choice of crop may be limited to flood-tolerant species such as rice and jute, a circumstance that may be dictated even in relatively short-season savannah climates. The cropping cycle may extend beyond a year, not merely through the use of perennials or semiperennials but by planting long-duration tuber crops that are harvested, sometimes intermittently, over a period of up to eighteen months.

■ 12.3.6 *Cropping tactics at high altitudes.* For any given rainfall regime, an increase in altitude is likely to have the effect of extending the growing season since temperatures are lower and evaporation is reduced. Moisture deposited from clouds may also be significant. In continental locations the growing season may be limited at either end by temperature, as in the foothills of the Himalayas. However, in equatorial high-altitude situations, where growing conditions are favorable year-round but the general range of temperature is low, crop species of tropical origin have evolved into very late-maturing types. Thus in the high valleys of Colombia, corn cultivars may take ten months or more to mature, though the ample moisture is utilized neatly by relay cropping with climbing beans after five months. Their early development is assisted by removing the lower nonfunctional corn leaves to allow more light to penetrate to the young beans.

■ 12.4 Agronomic tactics in relation to soil water

■ 12.4.1 *Competitive relations in mixed cropping, intercropping and relay cropping.* As discussed previously, crop monoculture in mechanized farming systems is dictated primarily by the limited capacity of farm machinery to cope, particularly at harvest, with more than one crop on one area. In nonmechanized tropical farming no such limitation applies and mixed cropping, intercropping, and relay cropping are at least as common as monoculture. Monoculture is associated particularly with cash cropping and with environmental situations that restrict the range of adapted crops. The advantages of complex cropping have already been discussed (see section 6.6) but not the biological and physical aspects of intercrop competition.

Although experimental work on the yield relations of monoculture and complex crop mixes has been in progress intermittently for some time, particularly in Africa, much greater attention has been paid to the topic in recent years. The results have been variable, but there is increasing evidence to suggest that with the appropriate crop population, planting geometry, plant habit, maturity type, and planting times, crop combinations can be found that yield more than the sum of the two crops grown separately in monoculture on the same given total area (e.g., Baker, 1974).

The dominant factor governing the productivity of the crop mix, except in situations where water is nonlimiting, is, as already discussed, the degree of coincidence between the seasonal pattern of combined crop water demand and soil water availability. In addition, as indicated in two examples referred to earlier, competition for light is of significance. In experimental work, nutrient competition may be overcome by adequacy of supply, but in the real-life situation intercrop nutrient competition, for nitrogen at least, is often diminished or eliminated by making one of the crop pairs a legume. The extended period of adequate crop cover in the intercrop situation may reduce soil erosion in comparison with monoculture.

At present, our appreciation of interspecific competition between annual crops is still in the empirical stage. The theory of interspecific competition in intimate mixtures of forage species, both annual and perennial, has been well explored, but we still await a comprehensive examination of the principles of competition between annual crops in the geometrical intercrop or relay crop pattern. A recent assessment of the situation has been made by Trenbath (1976).

■ 12.4.2 *Cultivation techniques for soil water conservation.* One of the major causes of inefficient water use in tropical annual cropping, other than that of the suboptimal matching of crop water demand to soil water availability, is loss of water as runoff occasioned by impermeable soil surfaces. We have already considered mulching as a means of soil conservation, but naturally its direct beneficial effects on crop yield, when they occur, are related to increased rainfall acceptance (and also to a reduction in soil evaporation and soil temperature). Another technique available to the farmer is timely cultivation. The example of soil water conservation in the black soil cropping pattern of India has been mentioned, though in this instance the cultivations given during the monsoon season are mainly for weed control and, if carried out when the soil is relatively wet, may result in compaction and reduced infiltration.

One of the critical periods during which useful water is lost as runoff is at the start of the wet season, before the land is ploughed. Experiments in a savannah climate in North Australia on a kaolinitic clay soil have shown the benefits of dry-season ploughing in increasing preplanting soil water storage and the yield of peanuts and sorghum (Phillips, 1959). Dry-season cultivation by hand or with ox-plough is often not feasible since the soil is too hard, but ploughing is possible immediately after harvest at the end of the previous wet season, while the profile is still moist, or in early dry-season showers (Krantz, 1974). In addition, many tropical soils develop a surface seal during crop growth: Lawes (1961) in Nigeria has demonstrated increased rainfall acceptance during the growing season with periodic hoeing. Finally, in this general category of cultivation techniques falls also the practice of tie-ridging; that is, the creation of basins to trap water by hoeing up ridges across the furrows (Lawes, 1961, 1966).

■ 12.4.3 *Ratooning.* Efficiency of water use by annual crops in semi-arid tropical conditions may be increased by the practice of ratooning, since neither time nor soil water is lost by land preparation for a repeat crop. In addition to the potential for achieving two grain harvests within a limited season, for example, of sorghum or millet, ratooning can also be used as an emergency technique for crop survival during a wet-season drought period. If a crop that has not reached maturity and is suffering from severe water stress is ratooned, and the forage fed to livestock, it may recover and produce a grain crop when the dry spell is broken (Krantz, 1974).

■ 12.4.4 *Determinacy of crop genotypes.* In the evolution of local crop genotypes, the selective pressure of a variable plant density (due to poor seed or seedbeds, pests, weeds, or diseases) and of the search for season-to-season yield stability has led to the development of indeterminate cultivars of flexible plant size. In an indeterminate genotype, the critical phases of flower and fruit development are spread over a long period and the probability of a single dry spell having a catastrophic effect on yield are reduced. There is some disadvantage in nonsynchronous maturity, but with hand methods successive harvests are possible. (With tuber crops of poor storage quality, whether in the ground or out, nonsynchronous maturity, providing food over a long period, is desirable.) Furthermore, in the event of a poor stand, individual plants are able to tiller or branch freely to utilize surplus soil water (and light and nutrients). Thus indeterminacy and flexible plant size are essentially hydrological adaptations. In modern tropical annual crop breeding, tillering or branching capacity is generally retained, but normally the aim is toward greater determinacy, which is associated with quick maturity, itself a major element in the strategy of plant breeding for improved water use in rainfed cropping systems.

REFERENCES

American Society of Agronomy. 1976. *Multiple cropping.*
ASA Special Publication no. 27.

Baker, E. F. I. 1974. *Research on mixed cropping with cere*
ing systems—a system for improvement. Hyderabad,
Workshop on Farming Systems. ICRISAT.

Bunting, A. H., and Curtis, D. L. 1968. Local adaptation of sorghum varieties in northern Nigeria. In: *Agroclimatological methods.* Proceedings of the Reading Symposium. Paris: UNESCO.

Cochemé, J., and Franquin, P. 1967. *An agroclimatology survey of a semi-arid area in Africa south of the Sahara.* World Meteorological Organization. Technical Note no. 86.

Fitzpatrick, E. A., and Nix, H. A. 1969. A model for simulating soil water regime in alternating fallow-crop systems. *Agricultural Meteorology* 6:303.

Indian Council of Agricultural Research. 1970. *Proceedings of the symposium on cropping patterns in India.* New Delhi, India: ICAR.

Kampen, J. 1974. *Soil and water conservation management in farming systems research for the semi-arid tropics.* Hyderabad, India: International Workshop on Farming Systems. ICRISAT.

Krantz, B. A., and Associates. 1974. *Cropping patterns for increasing and stabilizing agricultural production in the semi-arid tropics.* Hyderabad, India: International Workshop on Farming Systems. ICRISAT.

Lal, R. 1975. *Role of mulching techniques in tropical soil and water management.* IITA. Technical Bulletin no. 1. Ibadan, Nigeria.

Lawes, D. A. 1961. Rainfall conservation and the yield of cotton in northern Nigeria. *Empire Journal of Experimental Agriculture* 29:307.

Lawes, D. A. 1966. Rainfall conservation and the yields of sorghum and groundnuts in northern Nigeria. *Experimental Agriculture* 2:139.

Nye, P. H., and Greenland, D. J. 1960. *The soil under shifting cultivation.* Harpenden, England: Commonwealth Bureau of Soils. Technical Communication no. 51.

Okigbo, D. N., and Greenland, D. J. 1976. Intercropping systems in tropical Africa. In: American Society of Agronomy. *Multiple cropping.* Madison, Wisconsin: ASA Special Publication no. 27.

Pereira, H. C., and Hosegood, P. H. 1962. Suspended sediment and bed-load sampling in the Mbeya range catchments. *East African Agriculture and Forestry Journal* 27:123.

Pereira, H. C., Hosegood, P. H., and Dagg, M. 1967. Effects of tied ridges, terraces and grass leys on a lateritic soil in Kenya. *Experimental Agriculture* 3:89.

Phillips, L. J. 1959. *The influence of land preparation on the yield of peanuts, sorghum and cotton at Katherine, N.T.* Commonwealth Scientific and Industrial Research Organization, Australia. Division of Land Research and Regional Survey. Technical Paper no. 1.

Trenbath, B. R. 1976. Plant interactions in mixed crop communities. In: American Society of Agronomy. *Multiple cropping.* Madison, Wisconsin: ASA Special Publication no. 27.

Chapter 13

Irrigated Annual Cropping Systems: General Aspects

■ 13.1 Definition of irrigation and types of irrigation systems

■ 13.1.1 *Definition.* Whereas in arid or semi-arid climates, whether in the tropics or in temperate zones, the distinction between irrigated and nonirrigated crop production is fairly clear, the words "irrigation" and "irrigated" become increasingly difficult to define as we move toward the high-rainfall tropics, where the disposal of surplus water may be as important as or more important than the supply of additional water.

Ruthenberg (1971) defines irrigation as "those practices that are adopted to supply water to an area where crops are grown, so as to reduce the length and frequency of the periods in which a lack of soil moisture is the limiting factor to plant growth." While this definition has the merit of being specific, it is limited in its application to situations where demand exceeds supply at most or all times. Furthermore, in his classification of irrigation methods, Ruthenberg exceeds his definition by including the impounding of rainfall *in situ* in a wet rice field, which may be termed storage but can scarcely be called supply.

As an example of the difficulties, consider a Balinese hillside wholly terraced for wet rice production. The field at the top of the hill receives only rainfall, which is impounded within a bank to create the flooded situation desirable for wet rice, but by no definition can the crop be termed "irrigated." In periods of heavy rainfall, the surplus from this field drains into the one immediately below, and so on down the hillside. Is the second field "irrigated"? Certainly it receives water other than the rain that falls on it, but it is not a purposeful supply of additional water for crop growth but a drainage mechanism for the field above; and in turn its surplus is drained off into a lower field. The whole hillside could be designated an impounding system, or a drainage system, but not an irrigation system.

Perhaps the only satisfactory definition of irrigation, one that lacks specificity but is at least properly inclusive and exclusive, is *the controlled supply of water to a field, additional to the rain that falls on it, and the controlled disposal of surplus water from the field*, the critical word being "controlled."

Having provided a definition of irrigation, this is the appropriate place to remind ourselves that in this and subsequent chapters we shall be paying as much attention to uncontrolled water supply and drainage, in relation to wet rice production, as to irrigation.

■ 13.1.2 *Irrigation systems.* "Irrigation systems" is used here in a rather special sense. To the engineer, the phrase connotes methods of water storage, supply, and drainage; here it signifies type combinations of crop production system and method of water supply. With our more restricted definition of irrigation the classification is simple, particularly since sprinkler irrigation, a method in common use in advanced temperate agriculture, is scarcely used at all for annual crops in the tropics. There are three main categories, all of which may operate either on a small or large scale:

a. Gravity flow irrigation of upland, i.e., nonflooded crops.
b. Gravity flow irrigation of flooded crops, i.e., rice and jute.
c. Irrigation of flooded or nonflooded crops by water drawn up from wells, ponds or streams.

On a small scale, irrigation water for types (a) and (b) may be derived from farm or village storages, or by minor diversions, sometimes temporary, from rivers or streams. For the type (c) irrigation system, numerous methods are employed involving human or draft animal power and simple mechanical devices (e.g., the Persian wheel), but the use of electric or internal combustion motors for pumping is increasing rapidly. Examples of types of large-scale systems are:

a. The Gezira scheme in the Sudan, growing cotton, sorghum, lubia bean (*Lab-lab purpureus*), etc.;
b. Major river control systems in South and Southeast Asia, e.g.,

the Ganges, the Irrawaddy, etc., supplying water for rice, sugarcane, and upland crops;

c. Tubewell systems on the Indian subcontinent, supplying water for a wide range of crops.

■ 13.2 Advantages and disadvantages of irrigation

■ 13.2.1 *Advantages.* The basic advantages that accrue from the controlled supply of additional water to crops are these:

a. A major limiting factor to the effective use of solar energy and soil or fertilizer nutrients for crop growth is removed or diminished.

b. The crop-growing season is extended.

c. Season-to-season variability in yield is reduced.

Since a major yield limitation is lifted and the amplitude of seasonal yield fluctuation is reduced, it becomes potentially more profitable and more safe to increase other inputs (labor, fertilizer, pesticides, etc.) to obtain higher output. Extension of the growing season permits an increase in the number of crops grown each year (see Table 13.1), a broadening of the range of crops grown, and a more continuous supply of harvested food. In addition, the greater spread of production tasks over the year leads to more efficient use of a permanent labor force.

■ 13.2.2 *Disadvantages.* The main disadvantages of irrigation are these:

a. Irrigation requires an increased level of capital investment in storage, reticulation, and drainage structures. On a small scale, these costs may be borne by the farm unit itself or by a group of units. The capital costs of major irrigation structures are normally borne by the state, which distributes the burden over the whole population through taxation. Whether the operational costs of the system are borne by the state or by the user, through his paying for water, is a matter of regional or national policy.

b. Within the farm unit, maintenance of the reticulation and drainage system demands a substantial and recurrent labor input. The farmer and his labor force may also be required to give part of their time to similar tasks within a larger unit;

e.g., in communal maintenance of a village irrigation system.

c. The actual task of supplying irrigation water to crops, particularly when it has to be drawn up from wells and streams, demands an additional labor input above that of rainfed cropping.

d. The smooth operation of any irrigation supply or drainage system that extends beyond the boundaries of a single farm unit requires a degree of cooperation and exercise of authority within groups of farmers, or between farmers and a public authority. The communal maintenance work mentioned above is one example; an area of necessary cooperation in large-scale irrigation schemes is the giving of prior notice of water requirement and acceptance of one's place in a time roster of water supply.

TABLE 13.1
Changes in average cropping index in five Asian countries; 1950–1968

Year*	India	Pakistan	Bangladesh	Indonesia**	Taiwan
1950	1.11	1.11	1.28	—	1.70
1951	1.11	1.08	1.29	—	1.70
1952	1.11	1.07	1.31	—	1.72
1953	1.12	1.09	1.33	—	1.72
1954	1.13	1.12	1.31	—	1.74
1955	1.14	1.13	1.27	1.04	1.71
1956	1.13	1.12	1.27	1.10	1.76
1957	1.13	1.22	1.28	1.11	1.79
1958	1.15	1.14	1.27	1.21	1.80
1959	1.15	1.10	1.29	1.15	1.82
1960	1.15	1.07	1.32	1.20	1.84
1961	1.15	1.10	1.31	1.12	1.86
1962	1.15	1.09	1.32	1.22	1.85
1963	1.15	1.13	1.34	1.08	1.85
1964	1.15	1.15	1.35	1.26	1.88
1965	1.14	1.13	1.37	—	1.89
1966	1.14	1.12	1.38	—	1.90
1967	—	1.08	1.45	—	1.87
1968	—	—	1.39	—	1.88

SOURCE: Economic Research Service, USDA (1971).
*Where data are given for periods other than the calendar year, e.g., 1952/53, they are placed on the basis of the earlier year, i.e., 1952.
**Java and Madura only.

13.3 The availability of irrigation water

13.3.1 *Amount of available water or land.* For the individual farm unit, irrigation of all the crop area is normally only possible when the farm lies in a region served by a large-scale irrigation system, with large storages, a well-designed reticulation and drainage system, and with fields properly laid out for the full use of irrigation water. The farm layout may be created *de novo*, as when a dam is constructed or a tubewell system sunk and land not formerly cropped is brought into cultivation. If, on the other hand, an irrigation system is established in an area already developed for rainfed crop production, extensive leveling, relocation, and reparceling of farm units may be necessary.

Irrigation of only part of the farm area is far more common, the restrictions on whole-farm irrigation including perhaps the amount of water available, the proportion of land commandable, or the distribution of suitable soils. Partial farm irrigation may be found associated with both large- and small-scale schemes, but it is more characteristic of the latter. In general, irrigated farm units are smaller than rainfed farm units in the same environmental situation (see Kampen, 1974, and Table 5.2).

Where irrigation water is available in predominantly rainfed farming areas, there is normally a clear distinction between crops and cropfields that are fully irrigated and those that receive no irrigation water. Although timely supplementary irrigation to largely rainfed crops may give a greater response in crop yield per unit water applied than that obtained from fully irrigated crops, the limitations of topography, field layout, or water-raising facilities normally make it uneconomic. For the individual farmer with access to water, full irrigation of part of the farm is usually more profitable than supplementary irrigation of a larger proportion of it. Furthermore, supplementary irrigation is necessarily irregular and unpredictable, since it rectifies the deficiencies of irregular and unpredictable rainfall, and in large irrigation schemes there is rarely the required flexibility of supply at short notice.

The degree to which irrigation water is used efficiently will depend on two major factors: first, whether land or water is the greater limitation to output; second, whether the user has to pay for it and on what basis he has to pay. Thus in irrigation schemes like the Sudan Gezira, where water not land area is the main limitation, high water use efficiency is essential. In other areas, commandable land area may be more of a limitation than water supply; in such a situation efficiency in the use of water is less important. With respect to cost of water to the user, Thailand is an example of wasteful use: there water is traditionally regarded as a "free good" whether it comes from rivers or streams or from costly irrigation structures, and farmers have little economic incentive to make efficient use of it. However, charges for water are only effective in inducing efficiency of use when levied on a volume basis. In some Southeast Asian irrigation schemes where the individual farms are very small this is not feasible and charges are levied on the basis of area irrigated, which clearly provides no incentive for efficient use.

■ 13.3.2 *Type of crop grown under irrigation.* If we compare those crops that receive the bulk of irrigation water in the tropics with those that receive little, we find that the irrigated crops normally have one or more of the following characteristics:

a. They are cash crops: e.g., cotton in the Sudan Gezira.
b. They are high-value crops: e.g., vegetables. "High-value" here may mean high cash value or high nutritional value for family subsistence.
c. They are crops that respond well to a finely controlled water supply: e.g., rice.
d. They are crops that can only be grown because extension of the growing season through irrigation has permitted cropping in a lower temperature regime: e.g., temperate cereals such as wheat or barley.

■ 13.3.3 *Time for which water is available.* In tropical summer rainfall regions, the temporal pattern of irrigation water availability has marked effects on the character of cropping systems. Storages, both of surface water and ground water, are replenished during the wet season when there is a water surplus and irrigation may not be needed at all. However, the period for which irrigation water is available in the dry season, and the quantity available, will depend on the magnitude of the storage, the efficiency of reticulation if it is surface water and, in large multiple-use storage systems, on competition from nonagricultural end-users: e.g., for power, domestic and industrial use, desalinization of estuaries, etc. Except for nonagricultural competition, these limitations apply to small-scale schemes and to surface or groundwater irrigation, though deep wells are perhaps more reliable. The relative dependability of well irrigation is seen in data from the Punjab, India, where in districts irrigated by wells the average cropping index in 1968–69 was 1.5, compared with 1.25 in canal-irrigated districts. It is predicted that this difference in cropping intensity will widen in the future (Economic Research Service, USDA, 1971).

In savannah climate irrigation areas where storage is limited or reticulation inefficient, water may only be available for a short period in early dry season and only quick-maturing crops requiring little additional water can be taken. A longer period of availability will permit crops of longer duration or greater water demand to be grown. However, this applies only to irrigation systems where the catchment is geographically near the irrigation area and has a similar rainfall pattern. In very large river systems, the upper catchments of which may be a long way from the irrigation area, peak flood is delayed. Thus peak flow in the Mekong delta of South Vietnam is 2 to 3 months after peak rainfall in the area, since the river is over 2,000 km long and the upper catchment is in a different climate. The classic example is of course the Nile, which rises in the tropics in a region of high summer rainfall and flows north to be utilized in a warm-temperate Mediterranean climate.

■ 13.4 THE LABOR ECONOMY OF IRRIGATED CROPPING

■ 13.4.1 *Total labor input per unit area.* With increasing use of irrigation water, total labor input per hectare per year tends to rise. There are at least six possible reasons for this:

a. The operations of irrigation, particularly onerous in nonmechanized systems of groundwater irrigation or other situations where water has to be lifted.
b. The maintenance of supply channels and drains.
c. Extension of the cropping season and the consequent increase in cropping index, particularly important in savannah climates where under rainfed cropping labor use in the dry season may be very low.
d. Greater incidence of weeds hence a higher input for their control. This is less true for the difference between rainfed and irrigated wet rice than for that between rainfed and irrigated upland crops.
e. Higher marginal return to labor, as a consequence of higher

and more dependable yield potential and increased inputs other than labor, e.g., fertilizer.

f. Greater harvesting and processing inputs, since yields are higher.

Boserup (1965) gives good comparative data for rainfed and irrigated crop production in India, which will be examined in chapter 15. With reference to (c) above, there is a remarkably close correlation in Taiwan between cropping index and labor input over the past 50 years (Economic Research Service, USDA, 1971) (Table 13.2).

TABLE 13.2
Relation between cropping index and labor input. Taiwan

Period	Cropping index	Labor input ha^{-1}
1911–15 (base)	100	100
1931–35	113	114
1956–60	156	155

■ 13.4.2 *Seasonal labor demand.* Clearly the longer the potential cropping period the better the seasonal distribution of labor demand. (In addition, maintenance tasks on irrigation systems may be carried out in off-peak periods to increase efficiency of labor use.) An excellent illustration of this is given in Figure 5.1, which contrasts the following:

a. Rainfed wet rice cropping during the wet season in the Central Plains of Thailand. There are two labor peaks: land preparation, transplanting, and weeding from June to September and harvesting in December and January.

b. A two-crop system in the valleys of North Thailand where, with irrigation, tobacco, garlic, and peanuts are grown in the early dry season (January to April) after wet season rice (July to December). The only partially slack periods are late in the growth of rice (September–November) and after the harvest of dry season crops (May–June).

c. A multiple cropping system in Taiwan with a cropping index of 3, where two rice crops and a range of upland crops are grown under irrigation. Hired labor at the peaks, which are not extreme, helps to dampen further the seasonal variation in demand on the permanent labor force.

■ 13.4.3 *Sources of power on irrigated farms.* The use of power inputs other than human muscle varies greatly in irrigated cropping systems and is not necessarily related to the intensity of the system. Draft animals are a particular advantage in the heavy tasks of land preparation and puddling for rice and of lifting water; in farming systems of high cropping index, their speed in land preparation between crops is also advantageous.

On the other hand, intensive irrigated farming is often associated with dense rural populations and small farm areas very fully utilized for cropping. In such circumstances free range grazing for draft animals is scarce and may be limited to roadsides and canal banks; hence crop residues have to be carefully conserved or natural forage cut and brought from some distance. In low-lying rice areas in savannah climates there may be two periods of forage scarcity in the year: one in late dry season when nothing grows and one in mid to late wet season when much of the uncropped land is under water. In the delta of the Chao Phya river in Thailand, buffalo are housed in the wet season on raised wooden platforms with a thatched roof and are brought forage by canoe.

In regions of very high population density and small farm size, feed and labor resources just cannot be spared for draft animals and the farms may be worked wholly by hand labor, as they are in parts of Java, for example. An additional factor in intensive vegetable production systems near cities is the high value of land. Expensive land necessitates producing the utmost in cash crops and not diverting cropland or labor to the maintenance of inefficient converters like the ruminant.

Thus irrigation farming units at their most intensive may dispense with draft animal power. On the other hand, the potential returns from irrigated cash crop production are such that the main penetration of tropical agriculture by mechanical power has been into irrigated farm units: pumps and walking tractors on both large and small farms and large tractors on the larger farms. Often pumps are owned and tractors hired.

■ 13.5 Types of irrigated cropping systems

■ 13.5.1 Our categorization of rainfed cropping systems was based on cultivation frequency. We can recognize within irrigated annual cropping systems an analogous range of intensification, which we can denote by cropping index. These systems may be grouped into four main types:

a. One-crop systems.
b. Two-crop systems; i.e., with a cropping index on irrigated land of about 2.
c. More intensive systems still based largely on field crops. Such systems may have cropping indices up to 3.
d. Highly intensive vegetable-growing systems, which may have cropping indices up to 8.

■ 13.5.2 *One-crop systems.* In many of the drier semi-arid regions of the tropics, where rainfall is erratic and water storage limited, only a single irrigated crop can be grown. Thus rice farmers in Anantapur and Kurnool in India wait until their tanks are full before transplanting; after the crop has matured there is no water left for a second crop.

■ 13.5.3 *Two-crop systems.* As a small matter of nomenclature, the appropriate general term for systems with a cropping index of about 2 is "two-crop." The term "double-cropping" is usually reserved for the special case where the two crops are the same; e.g., double-cropping with rice. In savannah climates, the phrase "first crop" is applied to the wet-season crop and "second crop" describes the irrigated dry-season crop.

As indicated in section 13.3.1, the type of crop grown in the dry season and the method of growing it will be greatly dependent on the amount of water available for irrigation and the period for which it is available. This is illustrated below within the common context of a preceding wet-season rice crop. Harwood and Price (1976) have reviewed Asian multiple cropping systems based on rice.

Where the quantity of water is small and is liable to run out quickly, time is saved by planting directly into the rice stubble without cultivation; for example, with peanuts or soybeans in Indonesia. Yield is relatively low because of the lack of land preparation and because the flat rice bay is not an ideal medium for upland crops, since soil aeration is poor. However, inputs are correspondingly low. The crops planted are nearly always legumes, for there is little available nitrogen in an undisturbed rice stubble immediately after harvest.

Where the amount of water is greater and its period of availability longer, the land may be fully prepared after rice harvest for a dry-season upland crop. If it follows rice, the land is usually formed into beds or ridges for better drainage. Again, the second crop tends to be a legume: a flooded rice crop may be virtually independent of nitrogen, but an unfertilized nonlegume crop sandwiched between successive rice crops is still likely to be nitrogen-deficient, even with the oxidation of organic matter that follows ploughing and setting up beds or ridges.

However, there are many two-crop irrigation systems where the second crop is a nonlegume even when rice is grown in the wet season. The second crop may, for example, be a high-value cash crop to which it is profitable to apply fertilizer. The rice-tobacco system of North Thailand, where the tobacco is grown on raised beds mulched with straw from the preceding rice crops is such a system. In the technically advanced irrigated cropping systems of North India and Pakistan, wet-season rice may alternate with dry-season wheat or barley; both phases of the system utilize modern cultivars that receive substantial nitrogen fertilizer.

Finally, where plenty of water is available and winter temperatures are high enough, a second crop of rice may be grown in the dry season. The independence of the rice crop for nitrogen as a result of nonsymbiotic fixation makes this system feasible without nitrogen fertilizer (Walcott et al., 1977) though naturally yields are improved if it is applied, particularly to the dry-season crop.

■ 13.5.4 *More intensive field crop systems.* In a number of areas in the tropics and subtropics with ample irrigation water to supplement normal seasonal rainfall, and soils that are suitable for a range of crops, complex multiple cropping systems in which two or three field crops are grown each year are found. Perhaps the most well-known are those of subtropical Taiwan (Food and Fertilizer Technology Center for the Asian and Pacific Region, 1974). A common rotation is two crops of rice followed by upland crops, of which a wide range is grown. Another common sequence includes rice, sugarcane, and upland crops; the sugarcane occupies the ground for fifteen months but upland crops are interplanted in the cane in the early stages of its development. The sugarcane, which is itself relay-planted into the preceding rice crop before the latter is harvested, does not form a complete canopy until the interplanted upland crops have matured. A similar cropping pattern, slightly less intensive, is found in East Java, where a three-year rotation is followed (Table 13.3).

This rotation evolved under the Dutch colonial sugar estate system where each farm unit grew its own subsistence crops and provided the central mill with cane. It is highly labor-intensive, since in preparation for sugar the ricefields are hoed up by hand into rectangular raised beds separated by quite deep furrows, which are then hoed down again when the sugar is harvested. Although because of the extended sugar crop the

TABLE 13.3
Intensive irrigated cropping system. E. Java

Year	Months	Period (m)	Crop
1	Jan.-June	6	Rice
1–2	June-Dec.	up to 18	Sugarcane
2–3	Dec.-May	5	Rice
3	May-Dec.	6	Upland crops: soybeans, peanuts, sweet potatoes, etc.

cropping index of such a system is only 1.3, it is more intensive with respect to labor input per hectare than many systems with a cropping index of 2. Ruthenberg (1971) describes a series of intensive irrigated cropping systems in India, some including rice and others exclusively of upland crops. The proportion of the holding irrigated varies from 25 to 100 percent and the cropping index from 1.2 to 2.4.

■ 13.5.5 *Intensive irrigated vegetable-growing.* Tropical irrigated crop production reaches its ultimate peak of intensity in the vegetable farms of Southeast Asia, which are usually operated by Chinese, and where the cropping index may reach the astounding value of eight. The crops grown are nearly all leafy vegetables, which are raised as seedlings, transplanted into the main crop area, and normally harvested four to six weeks after transplanting. The new crop may be in the ground on the same day as the previous crop is harvested.

Fertility is maintained with purchased fertilizer and with manure, sometimes composted with vegetable residues from pigs and poultry maintained on the farm. Most of the work is done by hand, though rotary hoes for quick land preparation are becoming more common. Ruthenberg (1971) quotes one example from Singapore with a labor force of 8.3 adult male equivalents per hectare. On the other hand, as we shall see in chapter 15, nonhuman energy input in the form of mechanical power, fertilizer, and agricultural chemicals can be high (Newcombe, 1976).

■ 13.5.6 *Historical increases in cropping index.* The increase in intensity of cropping in regions with extensive irrigation facilities is illustrated in Table 13.1, which shows cropping indices from the early 1950s to 1968 in five Asian countries. The changes are not necessarily related directly to an increase in irrigation, but the two are closely associated.

■ 13.6 High-technology tropical irrigated cropping

■ 13.6.1 *The green revolution.* The main impact of the recent massive research and extension effort initiated at CIMMYT in Mexico and IRRI in the Philippines and subsequently developed at national centers, which has unfortunately acquired the label "green revolution," has been upon irrigated farming systems in summer-rainfall climates. The developments in the Philippines, Mexico, India, and Pakistan are too well-known to need any recapitulation here. The crops with which the greatest success has been achieved are wheat and rice; the core of the technology developed and transferred is one of fertilizer-responsive nonphotosensitive cultivars together with a "package" of improved agronomic practices.

The social consequences of this transfer of technology have been commented upon adversely and at times unthinkingly. This book, concerned as it is with biological and physical processes, is not the place to review the pros and cons of the argument. However, two important points deserve to be made.

First, the improved practices are inherently scale-independent, and every effort is made to develop and extend techniques that can be applied on the smallest farms. However, since in order to take maximum advantage of them easy access to technical information and credit or capital resources (for fertilizer, particularly) is needed, the larger and more progressive farms gain not merely a proportionate advantage, i.e., a greater total return by virtue of farm size, but also have a greater opportunity to implement the practices. Most national governments, through vigorous extension efforts and the provision of credit, are attempting to make the opportunities for implementation available to all.

Second, although early improved cultivars had limited disease resistance since the primary breeding aim was yield, the more recent varieties have resistance to a wide range of common diseases built into them. Furthermore, while the early cultivars (of rice) had limited tolerance of variable hydrological conditions, special types are being bred for a wide range of environments; e.g., for deep water, for resistance to salt, and for tolerance of drought. Hence it is no longer true to say that modern cultivars have a restricted environmental tolerance, thriving only in the absence of major pest and disease challenge, and only under good water control and high fertility conditions, and thereby requiring a high level of input. High fertility was never mandatory: if other environmental conditions are favorable but fertility is low the new cultivars normally yield as much as or a little more than traditional cultivars.

■ 13.6.2 *The interactions of inputs.* Hitherto in our discussion of tropical cropping systems we have been largely content to describe current farming situations with little reference to recent technological advances. This book sets out to describe tropical annual cropping systems as they are and not as they might become, and broadly speaking tropical rainfed cropping systems have benefited to only a minor degree from modern technology. This is not the case with irrigated tropical cropping systems, particularly those based on wheat and rice, and some brief consideration of technology transfer in irrigated crop production is called for at this juncture. Special emphasis is laid upon that entity with the sadly overworked epithet of "the technical package," which may be defined as a set of improved production practices that interact positively so that collectively they are more effective in increasing economic return than when implemented singly.

The elementary principles of the package can best be discussed in terms of a simple model of input interactions. Some of the more familiar of these include cultivar and nitrogen fertilizer, and nitrogen fertilizer and weed control. To use a fertilizer-responsive cultivar without increasing fertilizer input may be of little avail; to step up nitrogen application without a corresponding increase in weed control input may prove an actual disadvantage.

Let us assume that we have four technical inputs, each contrasted with its "control" or traditional practice, combined in a 2^4 factorial experiment on, say, rice, which is carried out a large number of times under a wide range of farm conditions. (This is a model version of what is in practice done.) Our treatment pairs are:

A: Traditional cultivar, not responsive to nitrogen; improved cultivar, responsive to nitrogen.

B: Traditional nitrogen fertilizer regime, i.e., nil; high nitrogen fertilizer application.

C: Traditional hand weeding; improved weed control practices, perhaps herbicides.

D: Traditional absence of crop protection measures; improved crop protection measures, perhaps pesticides or fungicides.

Let us assume that the maximum increase in economic return over traditional practice is obtained from the full package *ABCD*, and let it equal 100 units. In full, the package may be rather costly for the small farmer; furthermore, he may be very reluctant to change, at one time, four of his traditional practices.

Now let us consider a series in which the improved practices are introduced one at a time and cumulatively, e.g., *A*, *AB*, *ABC*, *ABCD*; or cultivar first, then cultivar plus fertilizer, then cultivar plus fertilizer plus better weed control, then all four. There are actually 24 such sequences; another might be *D*, *BD*, *BDA*, *BDAC*. Depending on how the four individual inputs, *A*, *B*, *C*, *D*, interact with each other, the increments (or possible decrements) in economic return when each is introduced cumulatively will differ.

Suppose, for example, that they interact to the maximum degree. This would mean that no single input, or pair of inputs, or even triad of inputs increases economic return, but that when all four are introduced together they produce the 100-unit economic response. That is, whether introduced as *A*, *AB*, *ABC*, *ABCD* or *B*, *BD*, *BDA*, *BDAC*, the pattern of successive economic return is 0, 0, 0, 100. This is an unfortunate situation, and happily a most unlikely one, but if it were the case there would be no choice but to attempt to "sell" farmers the whole package. Vigorous efforts would have to be made by extension facilities to transfer the entire complex of inputs, and credit arrangements would have to be made to make the complex of inputs accessible to the farmer.

In contrast, let us postulate a situation where one particular sequence, say *A*, *AC*, *ACB*, *ACBD* (it might well be otherwise), gives a successive increase in economic return of 25, 50, 75, 100 units. This might be regarded as ideal, in that the farmer could be introduced to the new practices singly and gain an appreciable increase in economic return with each successive step. The adjective "appreciable" is here meant literally: that is, an economic improvement clear enough to be appreciated by a traditionally minded and possibly mistrustful farmer.

We can be more subtle: since it is known that traditionally minded farmers, once they have been persuaded to accept one major technical innovation and have found it to their benefit, are more ready to accept others, one might hypothesize that the ideal series of successive economic responses is something like 40, 70, 90, 100 units.

■ 13.6.3 The above model of technology transfer is grossly simplistic, since the degree of interaction between the individual inputs of the package will vary according to the nature of the inputs, the physical environment and cropping system, and the existing technological level of farm operations. With respect to the nature of inputs, it might for example be possible to introduce, as a single component, a new cultivar of greater pest or disease resistance with little change in the farmer's methods or costs; provided, that is, that its agronomic requirements and product quality were close to those of the traditional cultivar that it was destined to replace, and provided the authorities made its seed cheap and readily available. On the other hand, with irrigated wheat or rice for instance, the inputs of water control, weed control, and nitrogen fertilizer level might interact strongly.

There is no need to elaborate on the variation in degree of interaction that might arise in response to differences in physical environment or cropping system, but the influence of existing levels of technology deserves comment since it generally acts adversely on the process of technology transfer. This is because, other things being equal, we can expect a high degree of input interaction when the level of technology is low, whereas when farming units are already operating at a moderately advanced level of technology further improvement can often be made by the acceptance of single new inputs or a partial "package." Yet normally it is the farmers operating at a low technological level who are least likely to be able to take advantage of new technology, either because of limited economic resources or more firmly rooted traditional attitudes.

There is no easy solution to the problem of introducing technical innovations to the traditionally minded or credit-limited farmer, except under authoritarian political regimes. Even then there are technical if not social problems. The purpose in presenting this simplified model is merely to clarify some of the technical problems. The word "technical" is stressed: the craft and science of agricultural technology transfer involves far more than mere assessment of the likely interactions of proposed inputs; it requires a deep understanding of the social, cultural, and economic motivations of the farmer and, on a national scale, of the politicians and bureaucrats as well.

REFERENCES

Boserup, E. 1965. *The conditions of agricultural growth.* Chicago: Aldine.

Economic Research Service, U. S. Department of Agriculture. 1971. *Survey of multiple cropping in less developed nations.* Foreign Agricultural Economic Report no. 91. Washington, D. C.

Food and Fertilizer Technology Center for the Asian and Pacific Region. 1974. *Multiple cropping systems in Taiwan.* Taipeh.

Harwood, R. R., and Price, E. C. 1976. Multiple cropping in tropical Asia. In: American Society of Agronomy. *Multiple cropping.* Madison, Wisconsin: ASA Special Publication no. 27.

Kampen, J. 1974. *Soil and water conservation and management in farming systems research for the semi-arid tropics.* Hyderabad, India: International Workshop on Farming Systems. ICRISAT.

Newcombe, K. 1976. Energy use in the Hong Kong food system. *Agro-Ecosystems* 2:253.

Ruthenberg, H. 1971. *Farming systems in the tropics.* Oxford: Clarendon Press.

Walcott, J. J., Chauviroj, M., Chinchest, A., Choticheuy, P., Ferraris, R., and Norman, B. W. 1977. Long-term productivity of intensive rice cropping systems on the Central Plain of Thailand. *Experimental Agriculture* 13:305.

Chapter 14

Irrigated Annual Cropping Systems: Hydrological Aspects

■ 14.1 WATER USE BY CROPS

■ 14.1.1 *Rice and upland crops.* The water requirement, that is, actual evapotranspiration, of individual crops is partly an inherent attribute of the species itself (largely a function of the duration of its growth cycle) and is partly governed by the evaporative conditions of the environment in which it is grown. Actual water use in the cropfield system, that is, the total amount of rainfall plus irrigation water that is lost from the crop root zone in the course of growth, in whatever form, is also influenced by the physical characteristics of the soil, particularly its drainage and runoff characteristics, and by the pattern of rainfall intensity and the quantity and timing of irrigation water.

Examples of average actual use for specific crops grown in specific seasons are given in Table 14.1. The data are from southern India (Achar, 1970).

TABLE 14.1
Average water use by specific crops

Crop (cultivar)	Season	Water use (mm)
Rice (IR 8)	Wet	1,550
Rice (TN–1)	Late dry	1,250
Cotton	Early dry	564
Wheat (Mexican)	Early dry	465
Pearl millet (HB–4)	Late dry	447
Sorghum (CSH–1)	Wet	374

Although the same crop grown in seasons of different evaporative demand may transpire different amounts of water, there is a tendency toward equalization under clear-sky dry-season conditions. High evaporative demand is associated with high temperature, which encourages quick maturation; hence evapotranspiration per day may be higher but the period for which water is required is less.

The figures quoted above show that water use by rice may be two to three times that of upland crops. The components of total water use, including seepage, percolation, and drainage outflow, are examined in more detail below; but one reason for the high values is the high evapotranspiration rate of a flooded rice field. For example, Kampen and Levine (1970) obtained the relations given in Table 14.2 between E_t and pan evaporation (E_p) in Laguna and Central Luzon, Philippines.

TABLE 14.2
Evapotranspiration relations of wet rice

Season	Growth stage	Relation
Wet	Vegetative	$E_t = 0.84E_p + 0.35$ mm day^{-1} ($r = 0.94$)
Wet	Reproductive	$E_t = 0.91E_p + 0.17$ mm day^{-1} ($r = 0.90$)
Dry	Vegetative	$E_t = 0.87E_p + 0.52$ mm day^{-1} ($r = 0.88$)
Dry	Reproductive	$E_t = 0.94E_p + 0.46$ mm day^{-1} ($r = 0.91$)

For upland crops at peak water demand, i.e., maximum leaf area, E_t is normally in the range of 0.8 to 0.9 E_p and at other times substantially lower.

■ 14.1.2 *Water use in wet rice cropping systems.* The water use pattern of wet rice in two contrasting seasons is illustrated by data given in Table 14.3 from eleven sites in Luzon, Philippines, in 1969–70 (IRRI, 1972).

TABLE 14.3
Water use pattern of wet rice

	Wet season (mm)	Dry season (mm)
Net irrigation	1,090	1,475
Rainfall	850	210
Drainage outflow	1,210	537
Net water use	730	1,148
Net use per day	6.5	10.2
Evaporation per day	3.4	5.6
Efficiency [net water use ÷ (net irrigation + rainfall)]	38%	68%

"Net water use" in Table 14.3 includes evapotranspiration, lateral seepage, and downward percolation. Net water use of wet-season rice was only 64 percent of that of dry-season rice. (In these tests average crop duration happened to be the same for both seasons—112 days.) However, efficiency of water use as defined was much less for the wet-season crop largely because of surplus drainage in periods of heavy rain. Whether this water is "wasted," in the sense of not contributing somewhere to crop growth, depends on its ultimate fate as it moves through the cropping area (see below). The height of the retaining banks has a potential influence on the amount of runoff, though if water is in substantial surplus due to rain, the limits to runoff will be set by the desired depth of water for the crop, assuming that surface drainage of the field is possible at the time.

■ 14.1.3 *Water use in wet rice irrigation areas.* The example examined in 14.1.2 refers to a single ricefield. In a larger irrigation area, efficiency of water use in the area as a whole will also depend on conveyance losses, that is, losses of water in the reticulation system before it reaches the field, and on the feasibility of reusing drainage water for irrigation elsewhere.

Conveyance losses between two points in an irrigation system will depend on the length of channels and their seepage and percolation characteristics. These vary widely, of course: in the study by Kampen and Levine (1970) referred to above, conveyance losses from seepage and percolation averaged about 10 percent of average channel flow. However, much of the lateral seepage finds its way into adjacent ricefields and is utilized for crop growth, hence reducing the actual losses to the system. The major factor contributing to low conveyance losses in rice-growing plains or deltas is the high water table. On the other hand, leakage and operational losses due to poor construction and poor water control may in specific instances be much greater than the general losses from seepage and percolation.

An example of the effect of reuse of drainage water is given in Table 14.4 from a 30 ha irrigation area in the upper Sampanga River Project, Philippines, over a period of 46 days in the wet season (Cerdan, 1973). Thus field efficiency was low and similar to that in Table 14.3 (38 percent), but according to Cerdan's estimates the recovery of drainage water for further use increased total project water use efficiency to nearly 50 percent. It may also be noted that of net water use, evapora-

TABLE 14.4
Effect of reuse of drainage water

a.	Inflow of irrigation water	436 mm
b.	Rainfall	463 mm
c.	Total water $(a + b)$	899 mm
d.	Drainage outflow	610 mm
e.	Net water use $(c - d)$	289 mm
f.	Evaporation	232 mm
g.	Percolation	47 mm
h.	Seepage	10 mm
i.	Estimated reuse of drainage	281 mm
	Field Efficiency (e/c)	32%
	Project Efficiency $(e/(c - i))$	47%

tion, percolation, and seepage accounted for 80, 16, and 4 percent respectively.

■ 14.1.4 *Water use in land preparation for wet rice.* A significant proportion of the water used in wet rice cropping is lost in the land preparation phase, which for transplanted rice includes the initial ploughing and subsequent puddling and smoothing. The amount of irrigation water required for these operations is dependent not only on rainfall during the fairly long period over which the operations are performed—up to eight weeks—but also on whether the land is being prepared after a dry fallow phase or following the harvest of a preceding crop, when the root zone is still moist and the soil surface little cracked. This is seen clearly in comparative data from IRRI (1972) given in Table 14.5.

Thus in the prior dry fallow situation 500 mm, or almost half the

TABLE 14.5
Water use in land preparation for wet rice

	Dry*	Moist**
Duration of land preparation	52 days	56 days
Net irrigation	851 mm	526 mm
Rainfall	289 mm	182 mm
Drainage outflow	417 mm	339 mm
Net use	723 mm	369 mm
Evaporation	223 mm	198 mm
Land soaking	500 mm	171 mm
Land soaking/(irrigation + rainfall)	44%	24%

*Means of 9 sites, prepared after dry fallow.
**Means of 5 sites, prepared after previous crop harvest.

total water—rainfall plus irrigation—was used to soak the dry and deeply cracked soil, whereas when the land was prepared after a previous crop the amount so used was reduced by almost two-thirds. Hence if irrigation water and rainfall are adequate and sufficiently well-distributed in time to permit continuous sequential cropping, they are likely to be used more efficiently than if they are in seasonally short supply and thus impose a dry, no-crop period.

■ 14.1.5 *Delay in transplanting wet rice.* In rainfed wet rice cropping systems in savannah climates, and also in irrigation areas of similar climate where storage is inadequate to supply water at the start of the wet season, transplanting of rice is delayed until the wet season is sufficiently advanced for the farmer to be assured of continuous rain or irrigation water to maintain the required depth of water in his fields. Depending on the rainfall pattern of the region, this delay may restrict the cropping system to a single crop each year, the type pattern of rainfed wet rice culture in relatively short monsoon or savannah climate regions.

If, however, a crop of rice is seeded directly rather than transplanted, it may be sown much earlier. In its early stages, until there is enough water to flood the field, the rice is essentially an upland crop. Provided the wet season is not too short, this may allow time to follow the first crop with a second transplanted crop. The procedure is illustrated in Figure 14.1. The main disadvantages of this system are the increased input for weed control in the early stages of the first crop and the fact that it has to be harvested in late wet season, which creates problems of grain drying.

It should be borne in mind that delay in planting, with its potential for restricting the number of crops grown each year, is not always due to the slow onset of the rainy season or to inadequate early-season irrigation water storage. In large irrigation schemes, farms near the source of supply may receive irrigation water well before those at the lower end of the system. Thus in the 5,700 ha Penaranda River Scheme in the Philippines, farmers in the upper part of the area can plant one month earlier than those in the lower part (IRRI, 1974a). However, such problems can sometimes be overcome by more equitable routing of water through the reticulation system. A further example of technically induced delay is when dry-season irrigation water is not available until a month or so after the end of the wet season, the interval permitting

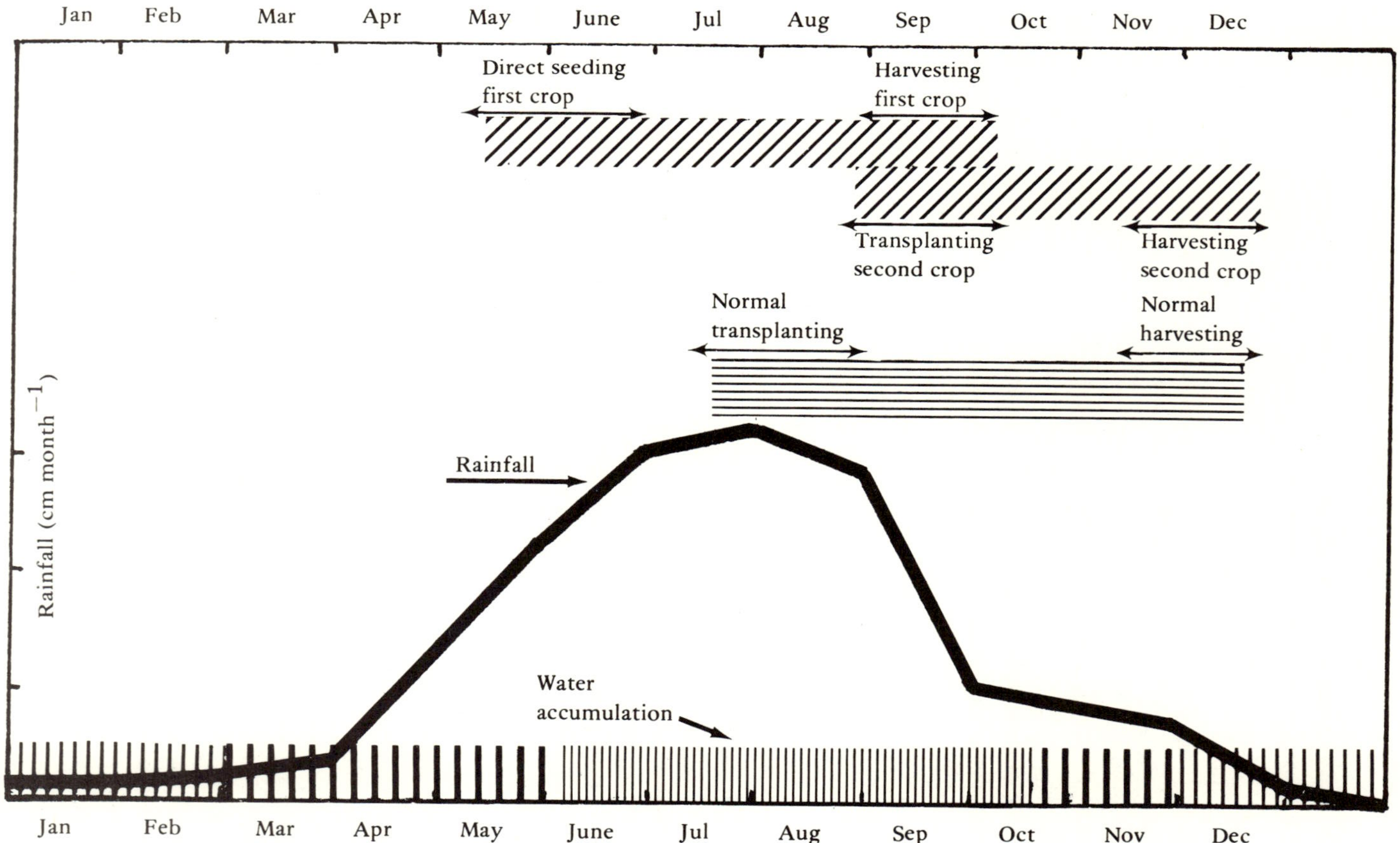

Figure 14.1. Utilization of rainfall by one-crop and two-crop rainfed wet rice systems in the Philippines (IRRI, 1975a).

maintenance work on supply channels and drains after wet-season damage.

Finally, social factors may be responsible for late planting. In rice cultures based primarily on a single wet-season crop, but where irrigation water is available for a second crop, the harvest of the first crop is all-important culturally. Its successful completion is followed by a period of festivity and general relaxation before the task of preparing land for a second crop is faced.

■ 14.1.6 *Wet rice and within-season water stress.* Wet rice is very sensitive to water stress during its growth period, and farmers go to great lengths in their attempt to maintain water levels. Figure 14.2 shows the response of IR8 rice receiving 100 kg ha^{-1} nitrogen to different intensities of water application over three dry seasons in the Philippines. In 1969, evaporative demand was higher than in 1970 and 1971; hence the rate of fall in rice yield with diminishing water supply was greater, though on the other hand maximum yield level with adequate water was higher. Attempts are being made at IRRI and elsewhere to select and breed cultivars tolerant of within-season water stress. The most direct way of countering this problem is to develop quick-maturing cultivars; i.e., stress-avoiding rather than stress-tolerant types. These may permit an increase in cropping index.

■ 14.2 Upland crop irrigation systems

■ 14.2.1 *Riverine "irrigation" systems.* In some tropical riverine situations, land submerged during the main period of wet-season river flow is planted as the water recedes to crops that grow to maturity on stored water. The same techniques may be used on flood-out areas away from the river banks.

By our definition such cropping operations cannot be termed irrigation systems, since there is little or no control over water. Their hydrological features do not differ in principle from those of a crop grown in a nonriverine situation on stored soil water derived from rainfall, though biogeochemically they may differ in that an annual deposition of riverine silt may maintain fertility. They are best represented on the banks of the Nile in the Sudan (Tothill, 1948), where wheat, barley, sorghum, lubia bean, and tobacco are grown. The banks of rivers in South and Southeast Asia are also cropped in this way: e.g., to peanuts and vegetables.

■ 14.2.2 *Basin irrigation.* In the arid northern Sudan, flooding and drainage of depression areas adjacent to the Nile have been controlled to the extent that the cropping pattern qualifies for the definition of an irrigation system. The practice is known as basin irrigation (Allan, 1948).

Supply channels are constructed from the river to the depression, together with one or more drainage channels back into the river. The basin may be divided by cross banks on the contour to hold up water, and peripheral banks may be built to keep the water within bounds.

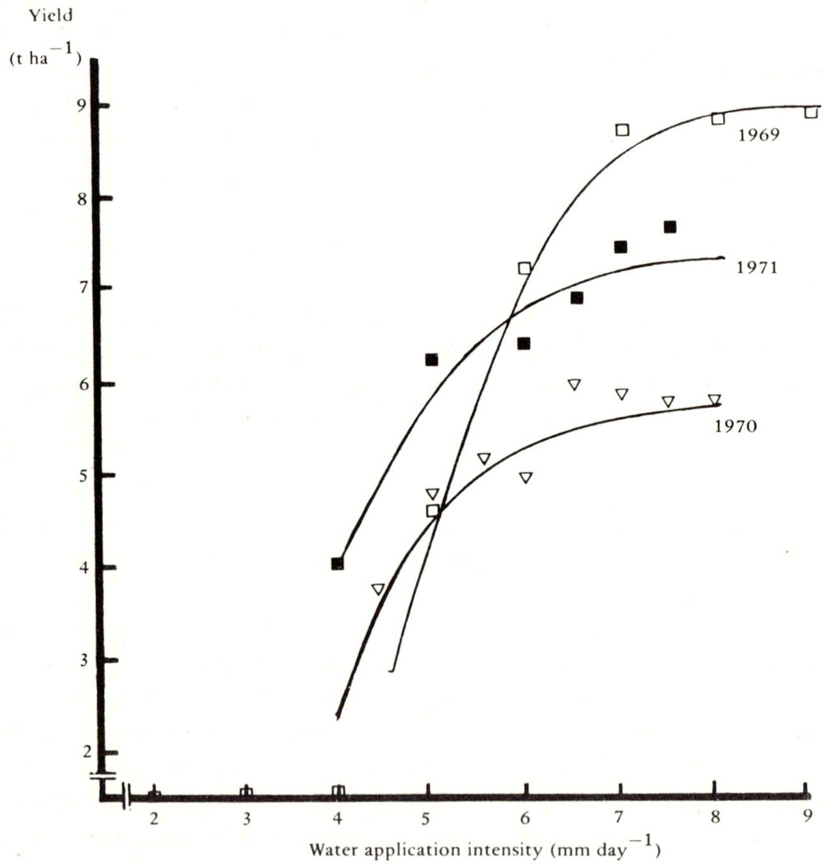

Figure 14.2. Response of IR8 rice to different intensities of water application over three dry seasons (Reyes, 1973).

When the river rises, the supply channel is opened and the basin flooded to the desired level, usually 70 to 100 cm; the water is held for about thirty days and then allowed to drain. If the basin is divided, the upper part may be flooded first and then drained into the lower part. (In Egypt, elaborate systems for moving water through a series of depressions have been developed.) When the basin has been drained it is cropped, e.g., with sorghum, chickpeas, etc. The reliability of flooding for any one basin depends naturally on its topographical position relative to the flood level of the river: the lower basins have a greater probability of receiving adequate water over a given number of years.

■ 14.2.3 *The Gezira scheme.* The oldest and largest of tropical crop irrigation schemes is that in the Sudan Gezira (Allan, 1948), which comprises over three-quarters of a million hectares. Rainfall in the area is low (200 to 400 mm), and water is used most efficiently. Formerly a cropping pattern of cotton, fallow, sorghum, and lubia bean was rigidly adhered to, and only 40 to 50 percent of the land was cropped each year, largely to conserve fertility. Cropping is now both more intensive and more diverse; peanuts and wheat, for example, are also grown.

The land is very flat and the soils heavy. The upland crops are irrigated by ridge and furrow methods, and the irrigation runs may be up to 280 m from channel to drain. Cotton is irrigated every 12 to 20 days according to evaporative conditions, and sorghum and lubia bean every 20 days. At the period of peak water requirement, about 7 mm day^{-1} of water is being delivered at the head of the main channel for crop irrigation; subsequent reticulation losses are 15 to 20 percent. The Gezira is a prime example of efficient tropical upland crop irrigation in an arid region.

■ 14.2.4 *Irrigated upland crop systems in monsoon climates.* Although irrigation farming on the Indian subcontinent is closely associated with wet rice cultivation, often in conjunction with other dry-season crops, there are types of irrigated annual cropping systems in which rice features little if at all. Ruthenberg (1971) summarizes some examples. Many different crops—cotton, pearl millet, peanuts, sorghum, wheat, etc.—are grown, and the patterns are difficult to categorize. They are mentioned here for completeness, and to note that frequently the farm units are only partially irrigated and that the crop area irrigated in the dry season often exceeds that given supplementary irrigation in the wet season. They will be considered in chapter 17, since many of them, whether they include wet rice or not, have the distinctive character of an irrigated forage component for ruminant livestock.

■ 14.3 IRRIGATED UPLAND CROPPING IN WET RICE SYSTEMS

■ 14.3.1 Type systems of multiple cropping with dry-season irrigated upland crops alternating with wet-season rice have already been reviewed in section 13.5. However, they have some specific hydrological features that merit comment, associated with the suitability or lack of suitability of the near-flat poorly drained ricefield for the growth of upland crops.

■ 14.3.2 In the situation where irrigation water is available only for a short period and in small quantities after the wet season has ended, second cropping is restricted to short-season upland crops. These are usually legumes such as peanuts, soybeans, field beans, etc. planted directly into the ricefield without land preparation, both to save time and to restrict inputs into what may prove a risky crop if water runs out before it matures. The technique is also used in wholly rainfed systems where, because of an extended rainy season or because soil water storage characteristics are favorable, the second crop can be brought to maturity on natural rainfall. The limitations to second crop yield in these situations are soil aeration and water availability. Of the legume crops, soybean is the most tolerant of poor aeration.

Seed of the second crop may be planted with a dibble in ricefields from which the straw has been removed, or planted or broadcast and allowed to establish through a surface mulch of rice straw. When the seed is dibbled, it is normally not planted at maximum distance from the old rice hills, but close to them, for two reasons. First, by the end of the rice-growing season the hill is slightly elevated above general field level; second, it is believed that the decaying root mass of the rice hill provides better aeration. The technique of planting or broadcasting seed into a rice stubble mulch reduces evaporation loss.

■ 14.3.3 Where irrigation water is available for a longer period but not in amounts adequate for a second crop of rice (rice usually being the preferred crop if conditions permit), upland crops can be given more favorable conditions for growth by modifying the geometry of the field surface. In fairly well-drained soils, which, however, may be hydrologically less efficient for growing the wet-season rice crop, major modification may not be necessary.

Various bed or ridge systems are in use, but in all cases the objective is to create a well-aerated medium for upland crop roots. The actual bed or ridge area on which the upland crop is planted is not inundated when irrigation water is supplied; essentially the crops are given subsurface irrigation, the water moving laterally from the intervening furrows into the beds or ridges.

Clearly the greater the number of furrows the higher the labor input, and hence beds are made as wide as possible in accordance with the lateral seepage characteristics of the soil. A width of 1.5 to 2 m is common. (In the Javanese rice–sugar upland crop system, the furrows for sugarcane may run in two dimensions, cutting the beds into small rectangles.) If the beds are high and the furrows deep, it is possible to hold water in them for long periods without detriment to the crop, which can develop an adequate root system wholly above water level. (In some delta areas of Southeast Asia, semiperennial and perennial crops, e.g., bananas, sugarcane, and even coconuts, are grown on high beds with permanent water in the furrows.) However, in heavy soils it may be necessary to allow water in only for the period required to saturate the bed and then to drain off the excess to avoid waterlogging. One of the advantages of the subsurface irrigation is that weed growth within crop rows is restricted since the soil surface remains relatively dry.

In intensive multiple cropping systems where upland crops are relay-planted into wet rice or into other upland crops, complex sequential patterns of ridging and bedding are followed in order to achieve smooth transitions from one desired hydrological situation to another.

■ 14.3.4 Finally, it should be appreciated that the irrigation of dry-season upland crops is intermittent, the intervals between irrigations for any one field being dependent on evaporative demand, crop type, and the drainage characteristics of the soil. Though actual water use by upland crops is substantially less than that by rice, large areas of upland crops can only be grown if an efficient time roster of water supply to individual users can be imposed.

■ 14.4 Spatial pattern of wet rice cropping

■ 14.4.1 In our examination of the spatial pattern of shifting cultivation and more intensive rainfed cropping systems it was possible to use as our unit of observation the individual farm. Factors such as topography and distance from the domicile were of importance, but these could be considered within the context of the single farm unit since, by and large, the environmental situation outside the farm or individual cropfield has no direct effect on the biological or physical character of the farm or cropfield itself. (Socioeconomic factors, however, such as distance from markets and from centers of alternative employment, may have marked effects.) On the other hand, in discussing the spatial organization of wet rice cropping, whether rainfed or irrigated, the hydrological aspects of the area outside the farm or cropfield are often most important since they govern the incidence and amount of run-on water and the feasibility of irrigation. The appropriate unit on which to base our appreciation thus becomes the catchment or subcatchment.

In this section, a limited series of type catchment situations, illustrated in Figure 14.3, is discussed. It is not claimed that the series is exhaustive; the examples merely represent some common situations in Africa and Asia. The influence of topography in concentrating runoff water for rice cropping shown by these examples reveals some of the limitations in classifying areas for rice cropping systems on the basis of climate alone (for example, IRRI, 1974b).

■ 14.4.2 *Mixed wet rice and upland cropping in West Africa.* Figure 14.3 (a) illustrates an organization of wet rice cultivation and upland cropping along the lower Gambia River that can be regarded as typical of West Africa (Ruthenberg, 1971). It represents a simple allocation of adapted crops to a gentle catena of slope and drainage within a single farm unit, with wet rice in the lower areas—in this instance protected from the brackish water of the estuary by a bank—and upland cropping on slightly higher land.

■ 14.4.3 *Asia: gentle topography, moderate rainfall, population not dense.* Figure 14.3 (b) is a typical cross section of the Northeast Thailand plateau. The upland areas are of poor sandy soils, often cultivated for rainfed peanuts, kenaf, etc., under irregular shifting or semi-intensive cultivation. Interdigitating with these gentle rises are shallow swales of somewhat heavier soils developed for rainfed wet rice. Although rainfall is of the order of 1,200 mm, between-season and within-season variability is high, and topography and soils are not conducive to efficient large-scale water storage. The swales are generally developed to ricefields to the limit of good seasons; that is, while ade-

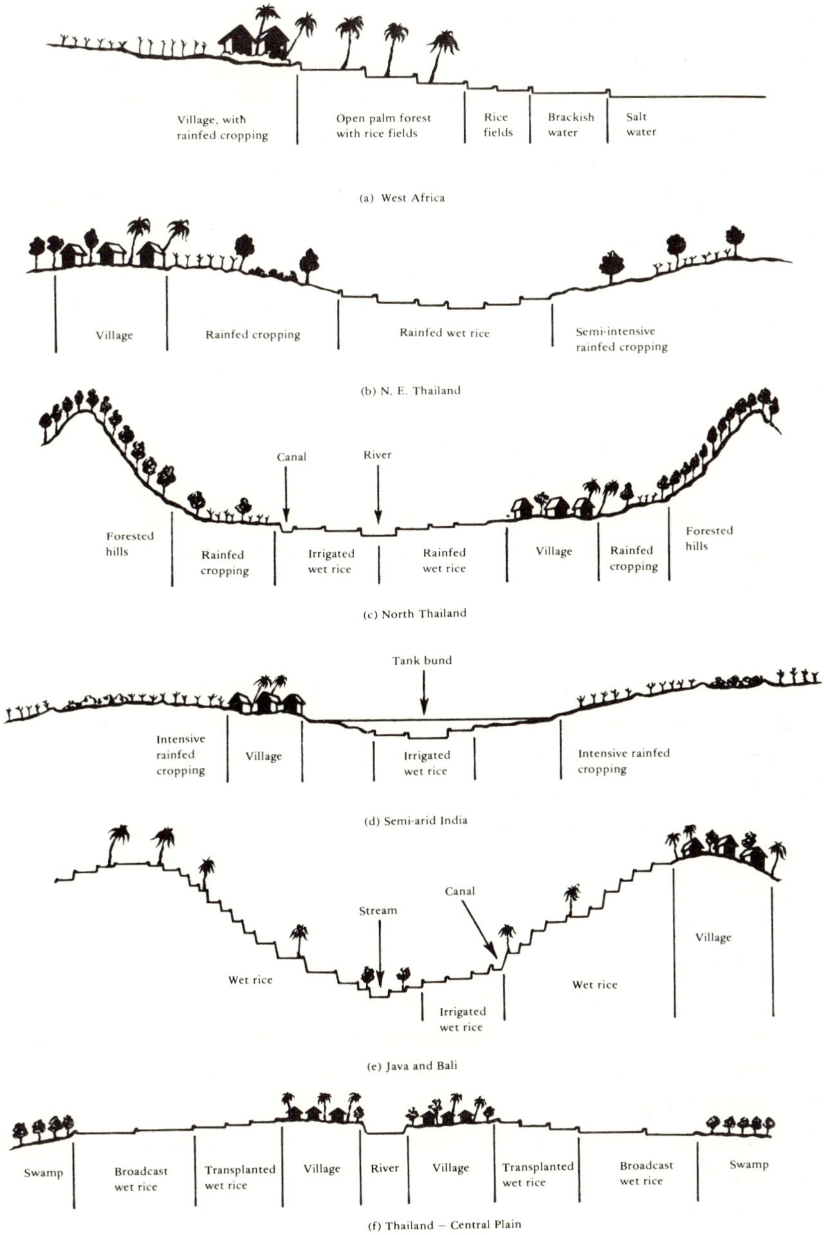

Figure 14.3. Spatial patterns of wet rice cropping.

quate crops may be taken from all of them in seasons of well-distributed and plentiful rain, in less favorable seasons crops in the upper fields may fail. In seasons where early rains are exceptionally poor the upper fields may not be planted at all. On the other hand, in years of unusually heavy rainfall the lowermost fields may be flooded too deeply for the type of rice cultivar planted.

■ 14.4.4 *Asia: abrupt topography, moderately high rainfall, population locally dense in valleys.* The situation depicted in Figure 14.3 (c) is common to many valleys in geologically young folded country in Southeast Asia. The tributaries of the Chao Phya River in North Thailand are a typical example. The narrow valleys, which may or may not have headwater storage for irrigation, are intensively developed for wet rice production and contrast sharply with the relatively unoccupied, forested hills. The limited area suitable for wet rice without extensive terracing may lead to heavy local pressure on cropland, with the result that farmers may be forced into supplementary cropping on the forested hillsides under shifting cultivation. This practice, if uncontrolled, is potentially destructive to the whole hydrological system. If between the valley floor and the steep hills there are areas of undulating upland, these may be developed for semi-intensive rainfed cropping. Where dry-season irrigation water is available for rice land, upland crops are grown, since the storages are rarely adequate for double cropping with rice. Groups of farmers may build temporary partial dams across rivers or streams, constructed of bamboo and brush wood, for small dry-season irrigation diversions. Swept away in the wet season floods, they are rebuilt each year.

■ 14.4.5 *Asia: gentle topography, low rainfall, dense population.* Figure 14.3 (d) is illustrative of many areas of semi-arid India, where gently rolling country is intensively utilized for upland crops, the runoff being stored in tanks and used for irrigated wet rice cultivation in the shallow valleys (Kampen, 1974). The area of rice may be only a very small proportion of the total cropped area of the region. As already mentioned, some of the tank systems have been in operation for many hundreds of years. In gently undulating country they are an inherently inefficient form of storage because of high surface-to-volume ratio, and with intensification of upland cropping and grazing in the catchment they have in many cases become even less efficient through siltation. In these semi-arid regions evaporative loss from ricefields, particularly in the "oasis" conditions created, is very high, and efficiency of water use in terms of total crop energy output from the catchment per unit of rainfall is low. With improved methods of soil conservation, water retention, and supplementary irrigation of the upland crops, total catchment productivity could be greatly increased.

■ 14.4.6 *Asia: rolling to abrupt topography, high rainfall, dense population.* Figure 14.3 (e) is typical of densely populated regions in Java, Bali, and certain parts of the Philippines where, in areas of fertile soil, human population density has climbed steadily to levels which demand total utilization of land, both hill and valley, for rice and other crops. Land stability is maintained by very careful terracing and attention to erosion from the steep banks between the fields. Small and often ingenious irrigation systems may be established in the valleys, but the upland areas are often not irrigated by the definition we have adopted. Cropping pattern is very dependent on the topographical situation and rainfall. In the drier zones at the tops of the hills it may be possible to grow only one crop of rice a year; lower down, rice followed by a legume, e.g., soybeans or peanuts, may be grown; and in the valley bottoms two crops of rice a year are often feasible. This refers to the cropfields within the banks: in addition, upland crops may be grown on the tops or sides of the banks since cropland is in such great demand. On steep slopes the crop area within the "field" may be no more than a meter wide, shorter than the vertical distance to the next field below, and in area no bigger than a bedroom closet. In the relatively level areas, where bank strength is not important, the banks between fields are shaved each year to a width that makes it difficult for the unskilled to walk along them.

■ 14.4.7 *Asia: deltaic areas liable to flooding.* The most complex pattern of wet rice environments is found in the deltas of the major rivers of South and Southeast Asia. Figure 14.3 (f) is a typical cross section of the Central Plain of Thailand, where the streams have built up levees of well-drained soil; though at the highest points in the transect, they may be only a meter or so above the lower areas. Human population is concentrated along the levees out of flood danger in linear villages, where the houses are surrounded by gardens in which subsistence vegetables and fruit are grown and livestock maintained. Unoccupied levee land is in demand for upland crops. Immediately away from the levee, where control of water is adequate since drainage is possible, transplanted rainfed or irrigated wet rice is grown in small fields. Further away from the river, in zones of greater liability to flooding, the hazards of cropping make the high-input transplanted rice system uneconomic; the fields become larger, and pregerminated rice seed is directly sown by broadcasting. Medium-sized tractors may be used for land preparation in place of the buffalo cultivation of the transplanted rice area. Weed control inputs and average crop yield are lower; the cultivars sown are late-maturing and have greater flood tolerance. At a further distance still from the streambed, or, to move in another dimension, further downstream in the same topographical situation, deep-water rice cultivars are sown (see section 14.6). Even more complex and subtle adaptations of rice-growing methods to the range of hydrological situations are found in the Mekong delta (Takaya, 1974).

■ 14.4.8 *Cropping intensity and water level.* Figures 14.3 (c) and 14.3 (f) illustrate one particular feature of low-lying wet rice areas in Asia. As one moves from a higher to a lower elevation, with limited water control, then up to a point the increased water supply may permit intensification of the cropping system—from a single crop of wet rice to wet rice plus an upland crop to three-crop systems. At the lower levels, however, drainage becomes so difficult that the cropping system is once more simplified to double-cropping with rice and, lower still, to a single wet-season deep-water rice crop. The latter phase of this catena is seen clearly for example, in the Brantas River floodplain in East Java near Surabaya. At the higher levels, intensive rice–sugarcane upland crop systems predominate, at intermediate levels the land is double-cropped with rice, and at the lower levels only a single crop of rice is grown, leading gradually to mangrove swamp and ocean.

■ 14.5 Temporal pattern of wet rice cropping

■ 14.5.1 Figure 14.4 represents a series of six Asian rice systems, illustrating the range of time patterns in wet rice cultivation in relation to rainfall, availability of irrigation water and winter temperature. They are all lowland locations. In all instances, rice is grown in the wet season, though in one case at least the term "wetter season" would perhaps be more appropriate. Wet-season rice is associated with various other cropping enterprises in accordance with the availability of water and with winter temperature.

■ 14.5.2 The locations depicted in Figure 14.4 are:

a. *Chainat, Central Plain of Thailand* (1,230 mm). Without irrigation in this savannah climate of moderate rainfall and a distinct dry season, only a single crop of rainfed wet rice is grown. Where irrigation water is available, the winter is warm enough for a second crop of rice or for "summer" upland crops.

b. *Cuttack, East India* (1,640 mm). In a summer rainfall climate of greater total rainfall, with a more extended rainy season and irrigation from the Mahandi Delta Irrigation System, double-cropping with rice is possible even though at this continental location winter temperature is rather low (14°C). Local flood-tolerant cultivars are grown in the wet season and modern cultivars in the dry season.

c. *Kelantan, Northeast West Malaysia* (3,360 mm). In this region the main rains, which are extremely heavy, come from the northeastern monsoon in the "winter" or low-sun period. With irrigation and some rain throughout the year, the area is double-cropped with rice. One of the problems of irrigation is flooding of the pumps in the very wet period.

d. *Leyte, Philippines* (2,430 mm). With high total rainfall, a long wet season and some rain throughout the year, together with irrigation, there is sufficient water and winter temperatures are high enough for three crops of rice to be grown in one year. Acceptance of modern rice technology has gone further in the Philippines than elsewhere, and in all three cropping phases modern nonphotosensitive cultivars are grown.

e. *Punjab, North India* (570 mm). This area has a semi-arid continental savannah climate. With tubewell irrigation there is enough water to grow rice in the wet season, but it is too cold in winter for rice (December and January, 4°C) and wheat is grown instead. The farmers of the region are progressive, and modern fertilizer-responsive cultivars of both rice and wheat are grown.

f. *East Java, Indonesia* (1,720 mm). In the Brantas floodplain of East Java, with moderately high rainfall, irrigation in the relatively dry winter, and warm winter temperatures, three crops may be grown each year: two crops of rice and, in the pre–wet-season period of minimum likelihood of irriga-

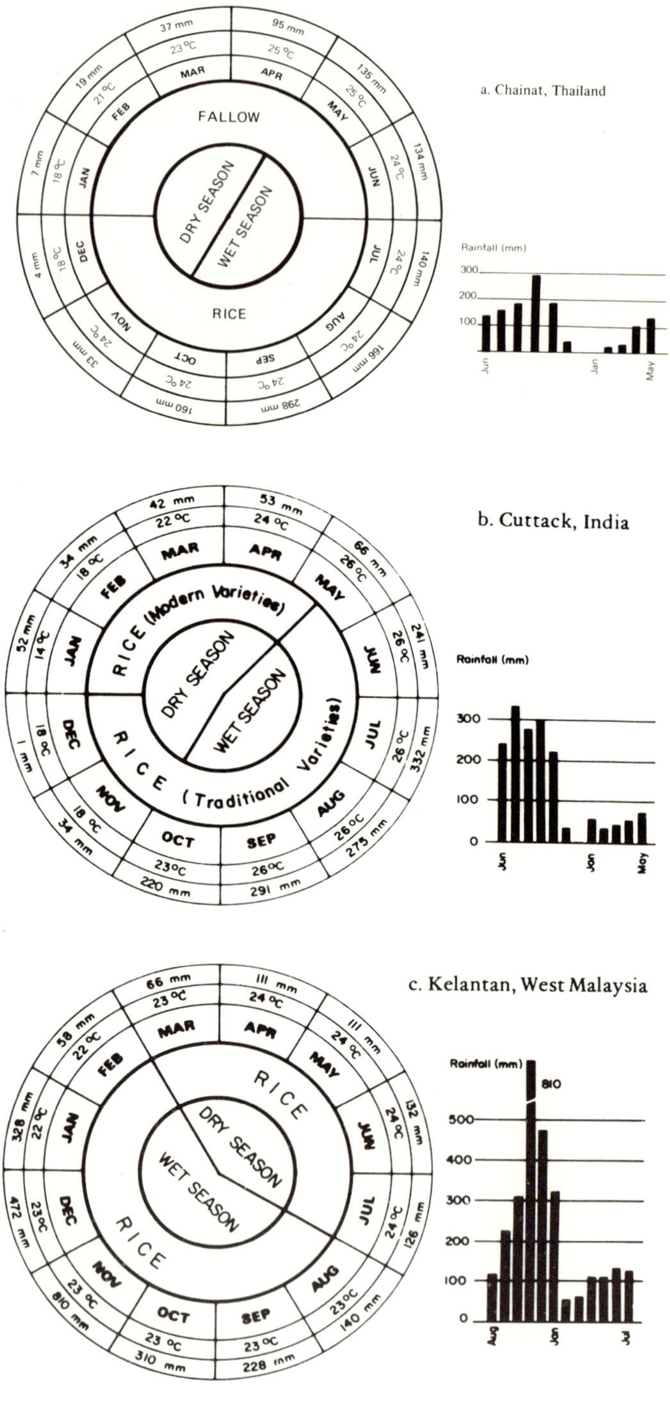

Figure 14.4a–c. Temporal patterns of wet rice cropping. (All except a from IRRI, 1975b.)

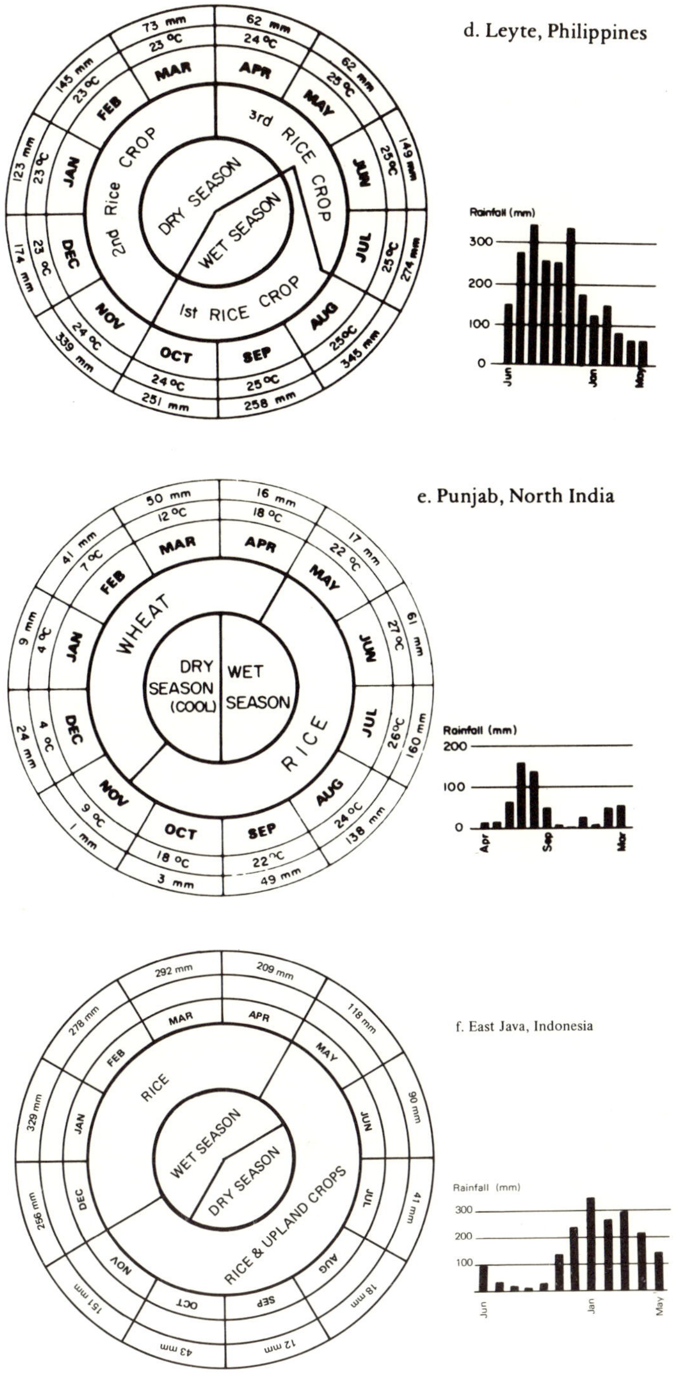

Figure 14.4d–f. Temporal patterns of wet rice cropping. (All except f from IRRI, 1975b.)

tion water, summer upland crops. In this area rice is often grown in association with sugarcane.

■ 14.6 Deep-water rice-growing systems

■ 14.6.1 Rice can be grown in areas where the maximum depth of water at the height of flood is up to 6 m. Such regions are characteristic of the deltas of the great rivers of South and Southeast Asia: the Ganges, Brahmaputra, Irrawaddy, Chao Phya, and Mekong rivers. (The equally long Salween River has very little commandable land, even at its mouth.) According to IRRI (1974c), something between 25 and 40 percent of the world's rice area floods to a depth of 1 m or more.

■ 14.6.2 The characteristics of the types of rice grown under such conditions include the capability to elongate rapidly as flood level rises and, for cultivars adapted to very deep water (more than 2 m), the additional capability of the leaves to float. It is estimated that of the 40 percent deep-water rice area, 10 percent is occupied by truly deep-water floating rice cultivars (2 to 6 m) and 25 to 30 percent by medium depth cultivars (1 to 2 m), which do not have floating leaves but which, like the floating rices, are capable of elongating rapidly. Elongation rates of 44 to 51 mm over five days for forty-day-old seedlings have been observed (IRRI, 1974c). These traditional adapted cultivars are all low-yielding, but they are being combined in breeding programs with modern high-yielding types.

In deep-water rice-growing systems, seed is broadcast at the start of the flood period on prepared land. The crops are late-maturing and display a local photoperiodic adaptation which brings them to harvest as the floods recede.

REFERENCES

Achar, H. P. 1970. Multiple cropping from available irrigation in southern India. *Indian Farming* October: 32.

Allan, W. N. 1948. Irrigation in the Sudan. In: Tothill, J. D., ed. *Agriculture in the Sudan*. London: Oxford University Press.

Cerdan, T. A. 1973. Water management plan for the upper Pampanga River project. In: *Water management in Philippine irrigation systems: Research and operations*. Los Banos, Philippines: IRRI.

International Rice Research Institute. 1972. *Annual report*. Los Banos, Philippines.

International Rice Research Institute. 1974a. *Annual report*. Los Banos, Philippines.

International Rice Research Institute. 1974b. *An agroclimatic classification for evaluating cropping systems potentials in Southeast Asian rice growing regions*. Los Banos, Philippines.

International Rice Research Institute. 1974c. *Annual report*. Los Banos, Philippines.

International Rice Research Institute. 1975a. *The* IRRI *Reporter* no. 5/75. Los Banos, Philippines.

International Rice Research Institute. 1975b. *Changes in rice farming in selected areas of Asia*. Los Banos, Philippines.

Kampen, J. 1974. *Soil and water conservation and management in farming systems research for the semi-arid tropics*. Hyderabad, India: International Workshop on Farming Systems. ICRISAT.

Kampen, J., and Levine, G. 1970. Water losses and water balance studies. *Philippine Agriculturalist* 54:283.

Reyes, R. D. 1973. Analysis of some factors affecting rice yield response. In: *Water management in Philippine irrigation systems: Research and operations*. Los Banos, Philippines: IRRI.

Ruthenberg, H. 1971. *Farming systems in the tropics*. Oxford: Clarendon Press.

Takaya, Y. 1974. A physiographic classification of rice land in the Mekong delta. *South East Asian Studies* 12:135.

Tothill, J. D. 1948. *Agriculture in the Sudan*. London: Oxford University Press.

Chapter 15

Irrigated Annual Cropping Systems: Biogeochemical and Energetic Aspects

■ 15.1 Biogeochemical aspects

■ 15.1.1 *Upland crop irrigation systems.* The supply of water additional to rainfall does not bring about any major changes in the biogeochemical principles of upland cropping systems. However, there are a few features of particular relevance to irrigated systems:

a. *Yield potential.* With the removal or minimization of one of the major factors limiting crop yield, that of inadequate water supply, yield ceiling is raised and season-to-season variation reduced. It therefore becomes potentially more profitable and economically less risky to apply substantial amounts of fertilizer, particularly nitrogen. No global data are available, but it is probable that well over 90 percent of the fertilizer applied to annual upland crops in the tropics goes to irrigated crops.
b. *Cropping index.* In order to maximize return on the investment in irrigation facilities, land with the potential to be irrigated is normally cropped intensively, at a cropping index of at least one. Hence in irrigated upland cropping systems, except on the most fertile soils, fertilizer inputs are normally required merely to sustain productivity.
c. *Leaching.* With a greater quantity of water reaching the soil surface than in rainfed cropping, the probability of losses of useful nutrients by leaching is increased on well-drained soils. On poorly drained soils, longer periods of saturation of the soil surface may increase denitrification losses.

d. *Salt accumulation.* On the other hand, in low-rainfall climates high evaporation rates and lack of thorough leaching may bring about an undesirable accumulation of salts in the crop root zone.

■ 15.1.2 *Salt accumulation.* In arid or semi-arid regions of irrigated upland cropping, salts may accumulate in the crop root zone to a concentration that adversely affects crop yield. Conditions favoring an accumulation of salts include:

a. High natural salt concentration in the groundwater.
b. High salt concentration in the irrigation water.
c. Inadequate application of irrigation water, so that little leaching occurs.
d. Hydrological conditions, e.g., a high water table, that retard or prevent leaching.

Even if the soils are of low natural salt concentration, salts may accumulate by deposition from irrigation water or by upward movement of groundwater, or both. The cations concerned are sodium and to a lesser extent calcium and magnesium; the important anions are chloride and sulphate.

If the soils can be leached successfully by periodic heavy irrigation then salinity in the crop root zone may be controlled. However, if leaching is not possible because of a high water table, then there may be a capillary flow of saline water upward into the crop root zone and deposition of salts at the point of evaporation. There have been instances of irrigation developments in the semi-arid tropics which, though successful at first, have become in time inoperative owing to the development of a high water table and the accumulation of salts. Crops differ in their ability to withstand salinity: cotton, for example, is fairly tolerant.

■ 15.1.3 *Wet rice systems.* The complex transformations and availability of nutrients in the flooded ricefield have been reviewed by Patrick and Mahapatra (1968) and Ponnamperuma (1972), and the nitrogen and phosphorus economy of wet rice has already been described briefly (chapter 4). However, in view of the increasing proportion of the world's wet rice area receiving fertilizer, particularly nitrogen, those aspects of rice nitrogen fertilizer nutrition relevant to cropping systems merit attention.

Readers will be fully aware of the development of stiff-strawed nitrogen-responsive photoinsensitive rice cultivars at IRRI and subsequently at other institutions, and the differences in nitrogen response between such cultivars and the locally evolved types that they are replacing requires no comment here. Two features of rice nitrogen response are, however, of relevance to cropping systems: the interactions of nitrogen and radiation and of nitrogen and water, illustrated in Figures 15.1 and 15.2.

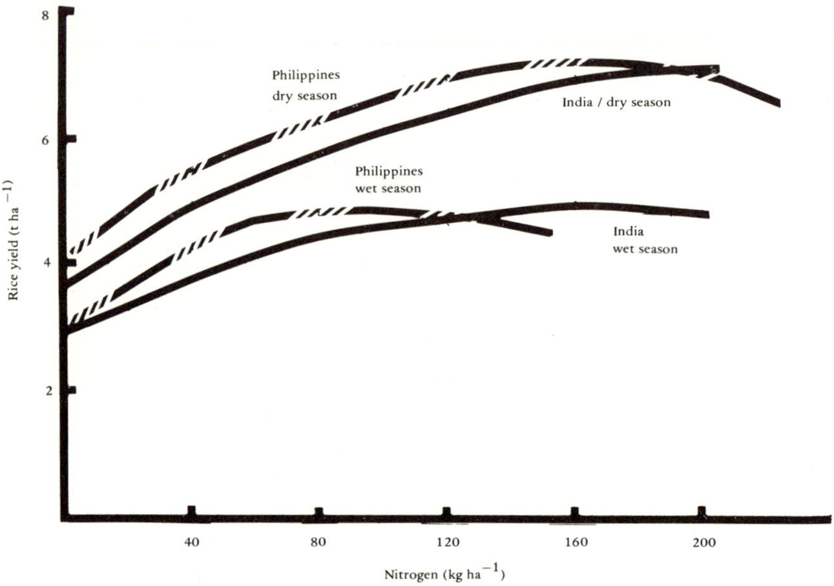

Figure 15.1. Response to nitrogen by rice in the wet and dry seasons in India and the Philippines (IRRI, 1969).

The average yield responses to nitrogen in the wet and dry seasons in India and the Philippines are compared in Figure 15.1. The pattern is similar in both regions: response to a given level of nitrogen fertilizer is greater in the dry season, with high daytime radiation and cool nights, than in the wet season. In many double-crop rice-growing areas, the proportion of dry-season rice area planted to modern cultivars is greater than that of the wet-season area so planted (IRRI, 1975). Modern cultivars are better adapted for dry-season cropping either because of the photoperiodic insuitability of local cultivars for dry-season cropping, or because the deep flooding expected in the wet season demands tall local cultivars. The most efficient use of fertilizer under such conditions is to apply all of it to the dry-season crop. In situations where modern cultivars are grown in both wet and dry seasons, more fertilizer is normally applied to the dry-season than to the wet-season crop.

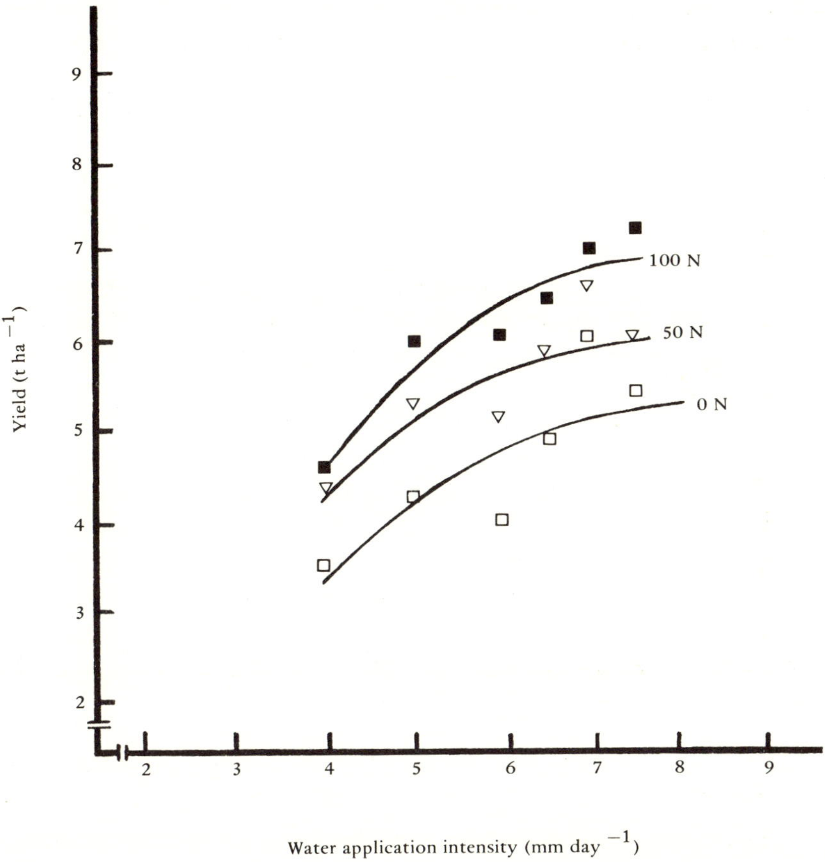

Figure 15.2. Relation between nitrogen input and intensity of water application for rice. Philippines (Reyes, 1973).

Figure 15.2 shows the relation between nitrogen input and intensity of water application at IRRI (mean of three modern cultivars). The figure illustrates the interaction of inputs referred to in chapter 13, in that unless the rice crop can be assured of adequate water, nitrogen response may be reduced. On the other hand, some modern cultivars such as IR5 do not show such marked reductions (Reyes, 1973).

■ 15.1.4 *Mixed wet rice–upland crop systems.* As already mentioned, dry-season cropping with irrigated upland crops within a summer wet rice system is usually restricted to legumes unless the farmer is prepared to apply nitrogen fertilizer to the second crop. Although in a wet rice field enough atmospheric nitrogen may be fixed to satisfy the needs of the crop, at least up to the limit of a 2 to 3 t ha^{-1} yield, the available nitrogen status of the soil between rice crops is low. Continued cropping with wet-season rice and dry-season upland nonlegumes without fertilizer may not only lead to low dry-season crop yields but may also depress the yield of the rice crops. On the other hand, continued cropping with wet-season rice and dry-season upland legumes is not likely to reduce rice yield.

■ 15.1.5 *Intensive vegetable-growing systems.* Although in intensive irrigated vegetable production in the tropics nutrient input was in the past largely from animal manure and compost, and from human excreta (now legislated against throughout the world on health grounds), increasing quantities of mineral fertilizer are now being used as a supplement to or a replacement for organic manure. Thus in Hong Kong the average annual rate of fertilizer application to cropland in 1971 was 1.1 t ha^{-1}, most of which was compound fertilizer, the remainder calcium nitrate and ammonium sulphate (Newcombe, 1976). The crop area was about half wet rice and half vegetables, but it is likely that the vegetable crops received the greater proportion of total fertilizer applied. Newcombe notes that the peak application to vegetable crops was nearly 9 t ha^{-1} yr^{-1}.

■ 15.2 Energetic aspects: comparison of rainfed and irrigated upland cropping

Table 15.1 gives some comparative figures from India of net human energy expenditure for upland crops grown under rainfed and irrigated conditions, converted from data in man-days collated by Boserup (1965), assuming an eight-hour working day and a net energy expenditure rate of 0.75 MJ hr^{-1}.

For all crop production tasks, energy input for irrigated cropping was on average about 2.5 times that for rainfed cropping. Land preparation, sowing, and harvesting inputs were increased by 36 to 45 percent, and inputs for weeding and manuring increased almost fivefold. (Weed growth in upland crops grown in the tropics with ample water is extremely vigorous.) The energy input for threshing was approximately doubled with irrigation, reflecting the higher yields obtained. The input for the task of irrigation alone averaged 28 percent of total input for irrigated crops and was equivalent to two-thirds of the average total input for rainfed cropping. Clearly the practice of irrigating upland crops requires a high additional energy input over rainfed cropping, particularly for the control of weeds and in the supply of irrigation water.

TABLE 15.1
Comparative net human energy inputs for rainfed and irrigated upland cropping. India (MJ ha^{-1})

Production task	Rainfed	Irrigated
Land preparation	53	77
Sowing	25	36
Manuring	1	6
Weeding	13	64
Irrigation	—	140
Harvesting	72	98
Threshing	38	74
Total	202	495

SOURCE: Boserup (1965); converted from man-days ha^{-1}, assuming 8 hr day^{-1} at 0.75 MJ hr^{-1}.

■ 15.3 ENERGETIC ASPECTS: WET RICE CROPPING

■ 15.3.1 Table 15.2 gives gross human energy input data for three contrasting wet rice cultivation systems: rainfed, gravity-irrigated and irrigated with water lifted by human and draft animal power. The input for watering has been excluded, and will be considered in the following section. In all three instances, rice was transplanted.

The table shows the general difference between the Thailand rainfed system and the more intensive Taiwanese and Indian irrigated systems, particularly in land preparation, weeding, and harvesting. Excluding watering, human inputs for the irrigated systems were approximately twice as great as for the rainfed system.

■ 15.3.2 The energetics of these three systems is further explored in Table 15.3, in which human and animal inputs and crop production and watering inputs are separated. Apart from energy expenditure rates, the only assumption that goes beyond the data (from

Ruthenberg, 1971) is that the 119 man-hours for land preparation in the Thai system also represents 238 animal-hours.

Line 3 shows the great difference in human energy input between irrigation by gravity and when water has to be lifted without mechanical pumps. Line 7 shows the corresponding difference for animal energy input. Lines 10 through 13 give the various output/net input ratios that can be calculated. Basing them on human energy excluding watering, the rainfed and gravity-irrigated systems are somewhat better than

TABLE 15.2
Gross human energy inputs for wet rice cropping excluding watering (MJ ha^{-1})

Task	Thailand (Rainfed)	Taiwan (Irrig.)*	Madras, India (Irrig.)**
Land preparation	119	206	376
Nurseries		115	
	185		280
Planting		112	
Fertilizing		83	130
	119		
Weeding		307	148
Harvesting	105	309	422
Other work	128	38	42
Total	656	1,170	1,398

SOURCE: Ruthenberg (1971); assuming energy expenditure 1 MJ hr^{-1}.
*Mean of spring and autumn crops.
**Irrigation water raised by human and animal power.

the water-lifting system. When human energy for watering is taken into account, it makes of course no difference to the rainfed system and little to the gravity irrigation system, but output/input ratio for the water-lifting system is almost halved.

When animal energy input is also taken into account, the output/input ratio of the rainfed Thailand system is reduced by two-thirds, and that of the Taiwanese gravity-irrigated system by half. For the Indian system output/input ratio is reduced to 1.7 because of the very high animal energy input for water-lifting.

■ 15.3.3 However, output/input ratios computed on the basis of human plus animal energy inputs for crop production tasks are not particularly relevant to human subsistence energy balance. Unless the animals are being fed on crop products that can also be consumed by humans and are thereby competing with humans for output, the important feature of the human energy balance is the energy expended by humans in feeding and caring for the livestock. As we have seen, this can vary greatly according to the amount of forage draft animals can gather with little or no human care.

TABLE 15.3
Energy inputs and outputs in wet rice cropping

	Thailand	Taiwan	India
1. Crop energy output (100 MJ ha^{-1})	196	325	258
2. Net human energy input, excluding watering (MJ ha^{-1})	492	878	1,048
3. Net human energy input for watering (MJ ha^{-1})	—	22	897
4. Total net human energy input (2+3) (MJ ha^{-1})	492	900	1,945
5. Net animal energy input for traction (MJ ha^{-1})	1,071	832	n.a.
6. Net animal energy input for watering (MJ ha^{-1})	—	—	n.a.
7. Total net animal energy input (5+6) (MJ ha^{-1})	1,071	832	13,293
8. Net human plus animal energy input, excluding watering (2+5) (MJ ha^{-1})	1,563	1,710	n.a.
9. Total net human plus animal energy input (4+7) (MJ ha^{-1})	1,563	1,732	15,238
10. Output/input ratio based on (2)	39.8	37.0	24.6
11. Output/input ratio based on (4)	39.8	36.1	13.3
12. Output/input ratio based on (8)	12.5	19.0	n.a.
13. Output/input ratio based on (9)	12.5	18.8	1.7

SOURCE: Adapted from Ruthenberg (1971); assuming net human energy expenditure 0.75 MJ hr^{-1}; net animal energy expenditure 4.5 MJ hr^{-1}; energy value of paddy rice 12 MJ kg^{-1}.

Data for the Thai system comes from the Central Plain, where cattle or buffalo have access to roadside and canal-side grazing, natural forage is cut and brought home, cattle graze rice stubbles in early dry season, and rice straw is saved for late dry-season feeding. The human energy input for animal care and feeding might represent an additional 10 percent above crop production tasks. On the other hand, in Madras it is likely that natural forage is less freely available and crop residues have to be carefully saved; an additional 30 percent above crop production tasks is perhaps appropriate. If we assume an additional 10 percent for the Thailand and Taiwan systems and 30 percent for the Indian system, then the output/net input ratios based on total human energy input, crop production plus animal care and feeding, would be approximately 36, 33, and 10 respectively.

■ 15.4 Energetic aspects: multiple cropping

■ 15.4.1 One of the most significant advantages of the intensification of annual cropping through an extension of the growing season with irrigation is a better seasonal distribution of labor. Not only is the distribution of labor input more even throughout the year (see Figure 5.1), but total labor input per year increases, thereby providing greater rural employment. For example, in Figure 5.1 gross human energy input for the Thai single-crop rainfed rice system was 1,700 MJ ha^{-1} yr^{-1}, while that for the Taiwan multiple cropping system, where two rice crops and an upland crop are grown each year under irrigation, the figure was 5,200 MJ ha^{-1} yr^{-1}, a threefold increase.

TABLE 15.4
Human energy inputs for wet rice production

Water Supply	Local cultivars	High yielding cultivars
Rainfed	100 (base)	98
Irrigated:		
Irrigation poor	115	113
Irrigation good	127	152
All farms	110	123

One aspect of cropping intensification in relation to labor use or human energy expenditure that has been a source of controversy is the effect of "green revolution" technology. As already discussed, though the technology may tend to become associated more with large mechanized farms merely because the larger farms accept it more readily owing to their greater capital and credit resources, the new practices are in themselves biologically and physically scale-independent. The introduction of high-yielding cultivars, and the associated increase in fertilization, weeding, harvesting, and threshing inputs may actually increase human energy input per hectare. The figures in Table 15.4, from Barker and Mangahas (1971), show the relative human energy inputs for wet rice production on 118 farms in Luzon and Laguno, Philippines, in the 1968 wet season. (The base is rainfed wet rice with local cultivars.)

■ 15.4.2 *Intensive irrigated vegetable cropping.* Energy inputs for intensive irrigated vegetable cropping in the tropics are very high indeed, for three main reasons:

a. Cropping index is high and farm work is in progress throughout the year.
b. The high-value cash crops repay a heavy human energy input per hectare.
c. Fossil fuel energy input in the form of mineral fertilizer is high.

For example, in the data given by Ruthenberg (1971) for a

TABLE 15.5
Human energy inputs for irrigated field and vegetable crops relative to wet rice cropping. Taiwan

Crop	Input
Rice	100
Potatoes (autumn)	78
Potatoes (winter)	106
Sugarcane (autumn)	145
Wheat	97
Peanuts	107
Rapeseed	80
Muskmelon	172
Mustard	207
Cabbage	256
Chinese cabbage	354
Tobacco	800

SOURCE: Economic Research Service, USDA (1971); original data from T. H. Lee.

Chinese vegetable farm in Singapore, the labor force was 8.26 adult male equivalents per hectare. Assuming an average (Chinese) workday of at least eight hours, this would mean an annual human energy input of about 24×10^3 MJ ha^{-1}. The amount of effort put into irrigated vegetable crops and upland field crops relative to wet rice cropping in Taiwan is shown in Table 15.5.

With increased application of mineral fertilizer and other agricultural chemicals for weed, insect, and disease control, and with increased mechanization, total energy inputs for irrigated vegetable production become higher still. Table 15.6 gives the energy inputs for vegetable-growing in Hong Kong, which *in toto* amount to 307×10^3 MJ ha^{-1}

(Newcombe, 1976). The human energy component is only about 3 percent of total energy input, the energy in fertilizer almost half and electrical energy input over one-third of the total. This degree of substitution of fossil fuel for human energy is comparable to that of modern U.S. corn production (Pimentel et al., 1973). In comparison with the more traditional Singapore system described by Ruthenberg (1971), the human labor force is reduced from eight to three adult male equivalents per hectare. Calculation of output/input ratio for the Hong Kong

TABLE 15.6
Energy inputs for modern intensive irrigated vegetable production. Hong Kong

	Inputs (10^3 MJ ha^{-1})	Percent of total
Herbicides, insecticides	4.4	1.4
Fertilizer	136.6	44.5
Machinery (1 rotary hoe)	30.3	9.9
Human labor (3 persons)	8.8	2.9
Irrigation (electric pumps)	114.3	37.2
Other (transport, seed, etc.)	12.4	4.0
Total	306.8	100.0

SOURCE: Newcombe (1976).

system is rather meaningless, since crop output is largely fresh leafy vegetables, the value of which for human nutrition bears little relation to energy content.

■ 15.5 ENERGETIC ASPECTS: MECHANIZATION

■ 15.5.1 The effect of increasing mechanization on irrigated field crop systems is illustrated in Table 15.7, which gives data derived from a survey in the Meerut district, North India. Singh and Chancellor (1975) surveyed twenty-six farms and placed them in six classes representing different degrees of replacement of human and animal power by electric or diesel motors and by tractors. The six classes were:

a. largely rainfed; human and animal energy only;
b. largely irrigated, by Persian wheel or canal; largely human and animal energy;
c. irrigated, with electric or diesel motors for pumping;
d. irrigated, with electric or diesel motors for pumping and other stationary farm tasks;

e. irrigated, with tractors for traction, transport, pumping, and stationary tasks; no electric or diesel motors;
f. irrigated, with tractors for traction and transport and electric or diesel motors for pumping and stationary tasks.

Table 15.7 shows that between categories (a) and (b), the difference being the proportion of crop area irrigated, farm size decreased but labor-to-land ratio and cropping index increased. Within the largely irrigated farms, increasing mechanization from classes (b) to (f) was associated with increased farm size and a reduction in human and draft

TABLE 15.7
Labor and land characteristics and work energy for six classes of farm with varying levels of mechanization. Meerut, India

	Class of farm*					
	(a)	(b)	(c)	(d)	(e)	(f)
No. of farms in each class	5	7	4	4	2	4
Farm size (ha)	4.8	2.1	3.0	7.0	11.0	9.7
Total no. of laborers	1.8	2.0	2.0	4.8	4.0	4.0
No. of draft animals	2.0	2.3	2.8	4.8	4.5	2.0
Laborers per hectare	0.38	1.05	0.67	0.68	0.36	0.41
Draft animals per hectare	0.42	1.10	0.93	0.68	0.41	0.21
Cropping index	1.0	1.6	1.6	1.7	1.7	1.6
Work energy (MJ ha^{-1})**						
Animal	347	1,764	1,199	677	133	252
Electric/Dlesel	—	371	2,653	3,874	—	3,773
Tractor	—	79	25	104	5,166	2,354

SOURCE: Singh and Chancellor (1975).
*For description of classes, see text.
**Animal-, motor-, or tractor-hours multiplied by estimated power output rating.

TABLE 15.8
Crop output from irrigated farms. Meerut, India

	Class of farm*			
	(b)	(c)	(d)	(f)
Wheat (kg ha^{-1})	2,549	2,378	3,902	3,343
Corn (kg ha^{-1})	1,186	2,040	1,005	1,568
Cotton (kg ha^{-1})	697	—	729	988
Rice (kg ha^{-1})	1,556	—	2,388	2,529

SOURCE: Singh and Chancellor (1975).
*For description of classes, see text.

animal labor per hectare. Cropping index remained roughly constant. The table shows the progressive increase in mechanical work energy input and corresponding decline in animal work energy input.

■ 15.5.2 The increase in total work energy input per hectare was not always accompanied by an increase in crop yield per hectare, as seen from the data given in Table 15.8. Singh and Chancellor give crop output only for farm classes (b), (c), (d), and (f).

In economic terms, the highest gross margins, taking all crops into account, appeared to be associated with farm classes (c) and (d); that is, those where electrical or diesel power only was used to replace human and draft animal power for stationary tasks. However, this cannot be regarded as a general truth; clearly the economic impact of introducing tractors will be very dependent on farm size and the relative costs of tractor fuel, diesel fuel, and electricity. The impression is gained that in this area farms were too small to take full economic advantage of tractor power.

REFERENCES

Barker, R., and Mangahas, M. 1971. *The impact of new technology on labor use in rice production.* Los Banos, Philippines: IRRI.

Boserup, E. 1965. *The conditions of agricultural growth.* Chicago: Aldine.

Economic Research Service, U. S. Department of Agriculture. 1971. *Survey of multiple cropping in less developed nations.* Foreign Agricultural Economic Report no. 91. Washington, D. C.

International Rice Research Institute. 1969. *Annual report.* Los Banos, Philippines.

International Rice Research Institute. 1975. *Changes in rice farming in selected areas of Asia.* Los Banos, Philippines.

Newcombe, K. 1976. Energy use in the Hong Kong food system. *Agro-Ecosystems* 2:253.

Patrick, W. H., and Mahapatra, I. C. 1968. Transformation and availability to rice of nitrogen and phosphorus in waterlogged soils. *Advances in Agronomy* 20:323.

Pimentel, D., Hurd, L. E., Bellotti, A. C., Forster, M. J., Oka, I. N., Sholes, O. D., and Whitman, R. J. 1973. Food production and the energy crisis. *Science* 182:443.

Ponnamperuma F. N. 1972. The chemistry of submerged soils. *Advances in Agronomy* 24:20.

Reyes, R. D. 1973. Analysis of some factors affecting rice yield response. In: *Water management in Philippine irrigation systems: Research and operations.* Los Banos, Philippines: IRRI.

Ruthenberg, H. 1971. *Farming systems in the tropics.* Oxford: Clarendon Press.

Singh, G., and Chancellor, W. 1975. Energy inputs and agricultural production under various regimes of mechanization in northern India. *Transactions of the American Society of Agricultural Engineers* 18:252.

Chapter 16

Mixed Systems of Annual and Perennial Crops

■ 16.1 Classification of types

■ 16.1.1 As we have seen in previous chapters, nearly all farming systems based largely on annual crops, whether rainfall or irrigated, include some component of perennial crops. Normally at the minimum some perennial fruit trees or vines are grown in the garden area around the domicile. Indeed, one of the best-known types of mixed annual and perennial cropping, the mixed garden or *pekarangan* complex of South and Southeast Asia, is but an extension of the home garden to the point where it represents a significant proportion of total farm area.

Exceptions to this generalization include rainfed cropping systems in marginal semi-arid climates where the dry season is too long for the satisfactory growth of perennial fruit trees, some high-altitude cropping patterns, e.g., in Ethiopia (Westphal, 1975), and some of the more isolated systems of shifting cultivation in wet forest where fruit and nuts are more likely to be gathered from wild than from planted trees; for example, in the Amazon basin. Even in the latter situation, however, the trees can be regarded as semicultivated since they may be deliberately spared when the forest is cleared for cropping with annuals.

In rainfall regimes that permit the husbandry of perennial crops as a major enterprise there is a very wide variety of cropping systems in which both annuals and perennials figure prominently. Annual and perennial crops may be grown together on the same land in various time and space configurations, or they may be on separate land areas. Both annual and perennial components may be subsistence crops, perennial subsistence crops being exemplified by the plantain and the breadfruit tree *(Artocarpus altilis),* or, more frequently perhaps, the annuals are subsistence and the perennials cash crops.

■ 16.1.2 *Classification of mixed annual and perennial cropping systems.* Mixed annual and perennial cropping systems not only exhibit a wide degree of variation but may in themselves be individually complex. Categorization is difficult and the classification given below displays some of the undesirable but inevitable features of the general typologies of farming systems discussed in chapter 1; it is pragmatic rather than strictly logical.

a. *Systems of annual crops with herbaceous perennials or semi-perennials.* For example: sugarcane-rice-upland crop systems of South and Southeast Asia; annual crops with sweet and nonsweet bananas in equatorial and subequatorial Africa; annual crops with sisal, mainly in tropical Africa and America.
b. *The "mixed garden" system of South and Southeast Asia.* Strictly, mixed gardens represent sectors only of complex garden-field crop systems, but the type is so well recognized that it merits a category of its own.
c. *Systems of annual crops with subsistence tree crops.* For example, annual crops with breadfruit, coconut, etc., commonly found in Oceania.
d. *Systems of annual crops with cash crop tree perennials.* For example, subsistence annual cropping in association with cocoa, coffee, coconut, rubber, or oil palm.
e. *Systems where annual crops are grown during the establishment phase of planted forest trees,* known as *taungya* systems or "agrisilviculture."

The limitations of the above grouping are immediately revealed when we attempt to place one very well-defined group of mixed annual and perennial cropping patterns, widespread in Uganda and Tanzania, in which bananas, coffee, and annual crops are grown. These could be classified under (a) or (d). Similarly, in annual crop–coconut systems, the perennial may be grown either for subsistence or as a cash crop and hence would appear under both (c) and (d).

■ 16.2 Advantages of a perennial component in the cropping system

■ 16.2.1 *Socioeconomic advantages.* The socioeconomic advantages of including a major perennial crop component in an annual-crop farming system are related to the life span of the crop and the fact that in many instances the perennial crop product is readily salable. Thus the development of an area of perennial crop represents an accumulation of income-producing capital, a form of semipermanent investment. Often the tasks of crop husbandry and harvest are less arduous than those of annual cropping, and perennials become an investment for old age or partial incapacity, or as a legacy to be passed to offspring. They also represent security against which money or goods can be borrowed. (This can lead to situations where individual trees are mortgaged or their ownership shared.)

The husbandry of perennial crops usually requires a less highly seasonal labor input than that for annual crops; hence perennials may be integrated readily into an annual cropping system. This integration is helped when the perennial crop is of a type the yield potential of which is not threatened if for some reason—low prices, illness, competing labor demand—it is periodically neglected or not harvested. A good example is rubber.

When the perennial is a cash crop the combination of an annual crop subsistence base and a permanent cash income-producer provides an economic stability, and the potential for an improved living standard, that may not be enjoyed by those who farm under a system based on annual subsistence crops or perennial cash crops alone. Furthermore, a perennial crop area rarely requiring urgent attention is a useful complement to nonagricultural employment.

Finally, as a further potential for advancement, if the small tropical farm with a perennial cash crop is located in a region of large-scale perennial cropping, then technical advances, arising primarily from local research financed by the plantation economy, are readily available to the progressive and enquiring farmer, whether they be new agronomic techniques or new genotypes. In addition, the small farm may be able to take advantage of the processing, transport, and marketing infrastructure established by the plantation economy. Lastly, in countries concentrating on specific perennial cash crops where small holdings are numerous, cooperatives supplying technology and marketing infrastructure may be established.

■ 16.2.2 *Biological and physical advantages.* The main biological and physical advantage in developing a substantial perennial crop component within an annual cropping system lies in the increased potential for exploiting the natural resources of the internally variable farming unit and for conserving these resources. Since with the establishment of a perennial crop cover the liability of land to erosion is reduced, areas within the farm unit unsuitable for annual cropping without a major labor input for terracing, etc., may be planted to perennials. One particular advantage of perennials is that they can be planted individually or in clumps in odd corners of the farm that are not readily cultivable.

Since the root systems of perennials can normally tap deeper layers than can those of annual crops, subsoil water and nutrient stores not available to annual crops may be drawn upon. Furthermore, in some instances, e.g., rubber, the offtake of plant nutrients on harvesting is small in relation to total nutrient turnover. Hence the maintenance of soil fertility is less of a problem than with annual cropping. This is by no means true for all perennial crops: for example, the continuous harvest of young foliage from tea constitutes a substantial drain on nutrient resources.

■ 16.3 ANNUAL CROPS WITH HERBACEOUS PERENNIALS OR SEMIPERENNIALS

■ 16.3.1 We can divide this category of mixed annual and perennial cropping systems into three subgroups: when the two types of crop are integrated in both time and space, that is, they are the components of a crop sequence; when they are merely integrated spatially, with annuals and perennials grown on the same land area but not in any particular sequence; and when they are not integrated and occupy different land areas on the farm unit.

■ 16.3.2 When annual and perennial crops are integrated in time and space the perennial is normally treated as a long-season annual or biennial within a relatively short cropping cycle. Typical examples are the sugarcane–rice–upland crop sequences of South and Southeast Asia. On the other hand, annuals and perennials may be combined in a less organized fashion: when an area that has been under an herbaceous perennial for some time is brought back into annual cultivation, as may occur with bananas and sugarcane.

The orderly short-cycle cropping sequences including sugarcane are nearly always associated with a well-organized milling infrastructure, the cane being a cash crop. On the other hand, when cane is grown largely for home consumption it usually occupies a small area on the farm semipermanently, perhaps for up to twenty years under fertile conditions. It may be planted as a block, in linear fashion or in some odd corner.

When sugarcane is grown in intensive cropping sequences, it does not necessarily occupy the land exclusively during its period of growth. In Taiwan, for example, cane in its early stages is intercropped with upland annuals that are harvested before the canopy of the main crop closes over (Food and Fertilizer Technology Center for the Asian and Pacific Region, 1974). Finally, where sugarcane or bananas are grown in shifting cultivation systems, crop duration is governed by the general length of the cropping phase, though the perennials may survive in the recovering forest for a few years after annual cropping has ended.

■ 16.3.3 Throughout many regions of the tropics where the dry season is relatively short, sweet and nonsweet bananas are grown as the main source of carbohydrate in association with annual crops. The temporal and spatial relations between the bananas and the annual crops vary greatly. Bananas may merely occupy odd corners of the farm more or less permanently; they may be integrated with annuals in continuing intercropping patterns on the same land area; or they may be in pure stands separate from the annual crops. In the last situation, bananas are likely to be intercropped with annuals during their early development phases before they form a complete canopy.

A typical example is the banana–finger millet (*Eleusine coracana*)–cotton system of Uganda, which is found bordering the more extensive banana–coffee–annual crop region in climatic zones where the dry season is too long for coffee (Parsons, 1970). In this area the banana is near the margin of its climatic adaptation, but for the Bantu people who live in Uganda it is the preferred staple carbohydrate, a situation analogous to rice-growing on the northeastern plateau of Thailand. In even drier areas bananas are still grown but cassava replaces finger millet as the secondary energy crop.

■ 16.3.4 When sisal is grown on small tropical farms, a practice confined largely to tropical America, the capacity of the crop to deplete soil water is such that it is normally grown as a pure stand or in linear fashion as a hedge. Grown as a hedge plant it may function both as a cash crop and as a means of keeping livestock out of cropfields.

■ 16.4 The mixed gardens of South and Southeast Asia

■ 16.4.1 In lowland regions of South and Southeast Asia with an extended growing season, in association with wet rice or rainfed upland cropping, the home garden component of the farm unit may assume considerable significance and can no longer be regarded as a mere adjunct to field cropping. Thus in Java in 1940, mixed gardens accounted for 18.7 percent of the total area under cultivation, and in Central Java 22 percent (Terra, 1954).

The descriptive term "mixed garden" was coined by Willis (1914) to describe the "wild jungle-like mixture of fruit trees, bamboos, vegetables, etc." characteristic of the cropping pattern. The mixed garden is a largely perennial analog of the complex annual crop mixes found in shifting cultivation systems of the wet tropics. Terra (1954), who gives a detailed account of the mixed garden or *pekarangan*, observes that this type of crop complex is also important in Sri Lanka, on the Malabar coast of India, in parts of Burma, Thailand, Malaysia, and Indo-China, and in Polynesia, Micronesia, and the West Indies. Where necessary and possible, the gardens are irrigated, often very ingeniously with small channels, bamboo pipes and gutters, etc.

■ 16.4.2 Where mixed gardens are associated with wet rice cultivation the garden areas and homesteads are confined to islands of higher ground in a sea of paddy fields. In drier areas, e.g., East Java, they tend to occupy the better-watered and more fertile locations within a rainfed upland crop landscape.

The total productivity of the mixed garden in terms of energy and protein yield may or may not exceed that of the open arable fields, but its nutritional value is measured more in terms of the variety of food produced and in vitamin content. Furthermore, the garden yields "industrial" products such as bamboo, for all construction purposes, and palm leaves for thatching. Terra notes that with increasing population density in Java, the proportion of cultivated land under mixed gardens tends to increase. Ochse and Terra (1934) estimated energy and protein production in Java from *pekarangans*, *sawahs* (wet rice land growing rice, soybeans, and peanuts) and *tegalans* (rainfed upland crop areas growing mainly cassava and corn, with some legumes), and obtained the figures in Table 16.1.

TABLE 16.1
Estimated energy and protein production from different types of cropland. Java

Type of cropland	Energy output (MJ $ha^{-1} \times 10^3$)	Protein output (kg ha^{-1})
Pekarangan	18.8	58.5
Sawah	21.6	113.5
Tegalan	17.4	53.0

■ 16.4.3 The number of crop species grown in the mixed garden is very large and the crop canopy multilayered; full utilization of radiant energy is thereby achieved. Terra (1954) lists fifty-seven fruit trees, seven nut trees, thirty-five "vegetable" trees (i.e., those yielding nonsweet fruits, edible leaves, etc.), and six spice trees. The understory of annual vegetables and fruits listed includes forty-five types, and in some instances only genera are named.

■ 16.5 Annual crops with subsistence tree crops

■ 16.5.1 Terra (1954, 1958) has reviewed the historical development of the mixed garden and the culture patterns associated with it. The cropping pattern may be considered as a complex evolution from early forms of Southeast Asian and Polynesian cultivation systems based on tuber and tree crops (yam, coconut, breadfruit, etc.), which were prevalent before the spread of rice and other seed crops from China and India (Grigg, 1974). The cropping complex was gradually enriched by the incorporation of species from other regions, particularly from the New World via the Philippines: for example, sweet potatoes, cassava, field beans, taro or tannia (*Xanthosoma sagittifolium*), etc.

■ 16.5.2 Thus the somewhat simpler mixed annual and perennial subsistence farming systems of the Pacific Islands may be regarded as archetypal. The main tree species are breadfruit and coconut, the latter having many uses; the annual species include taro, yams, and sweet potatoes. A broad distinction can be drawn between the somewhat impoverished cropping systems of the infertile low atolls, where the crop mix may be restricted largely to coconuts and pandanus on sandy well-drained areas and taro in wet locations, and the richer systems of the high-fertility volcanic islands where, in addition to the above, a wide range of annuals and perennials can be grown, including sweet potatoes, yams, breadfruit, citrus, etc. (Barrau, 1961). In both instances, plant energy sources are supplemented by animal protein from fish, pigs, and poultry.

■ 16.6 Annual crops with cash crop tree perennials

■ 16.6.1 All the main tree cash crops that form the basis of plantation systems are also grown on small tropical farms that include a subsistence annual or perennial crop component. Of the six main crops—coconut, rubber, oil palm, tea, coffee, and cocoa—tea is the one confined largely to plantations, since the time from harvesting to processing must be short and processing is not readily accomplished on a small scale. Oil palm production is even more critical with respect to time between harvesting and processing, but processing can be done on a small scale and the product marketed in small quantities locally. However, smallholder tea-growing is found in Sri Lanka and Taiwan and is receiving governmental encouragement in Kenya. Ruthenberg (1971) gives an excellent series of case studies of small-scale tree cash crop production.

■ 16.6.2 *Development pathways.* The establishment of areas of perennial cash crops on small tropical farms does not stem necessarily from a decision to plant a designated area at a designated time. We can recognize four main development pathways:

a. *Planned establishment.* Smallholder tree crop farm units may be established *de novo* in a planned development scheme, normally under government auspices. Land areas are allocated or leased for the tree crops, with or without a family subsistence crop area. Smallholder rubber in Malaysia and tea in Kenya are typical examples. In the more advanced projects, the small farm units are integrated with a nuclear estate, which provides the processing, transport, and marketing infrastructure and acts as a center for technology transfer.

b. *Subdivision of plantations.* In some countries, because of political decisions on the social desirability of small-scale production, large tree crop plantations have been subdivided and leased or made over to small farmers. Coffee in Brazil is an example. The existing processing, transport, and marketing infrastructure may be retained for the benefit of the smallholders.

c. *Systematization of wild crop gathering.* In areas where important cash crop perennials grow naturally, e.g., coffee in Ethiopia, oilpalm in West Africa, farming systems have evolved from a starting point of wild crop gathering. The ratio of crop trees to other trees is gradually increased by planting crop seedlings and thinning the forest. Whether this process is continued to the stage of a pure crop stand is partially dependent on the desirability of retaining forest trees for shade.

d. *Perennialization.* This rather cumbersome noun is used here to name a process under which the perennial cash crop element in a mixed annual and perennial system gradually becomes the dominant partner. Thus in West Africa, cocoa may begin as a domicile garden plant, annual crops being grown for subsistence under shifting cultivation or semipermanent rainfed cropping. Gradually the area planted to cocoa is increased by interplanting in the field crop areas which, instead of reverting to forest, become perennial cash crop stands. Such areas are not necessarily permanent: a decline in productivity through disease or lower fertility may cause cocoa land to be abandoned for new forest areas.

■ 16.6.3 *Examples of annual crop–cash crop tree perennial systems.* The following brief examples illustrate the wide range of farming systems that fall into this category:

a. *Banana–coffee–annual crops.* In high-rainfall areas of upland East Africa a stable mixed cropping system dominated by plantains and coffee has evolved. (Plantains are frequently planted as shade crops in young coffee areas in tropical America). The annual crops grown include corn, sweet potatoes, peanuts, and vegetables (Parsons, 1970). A feature of these systems is the multilayer crop mixture around the domicile: bananas may overtop coffee, coffee overtops corn, which in turn overtops low-growing vegetables (see Ruthenberg, 1971).

b. *Cocoa and annual crops.* (West Africa). The basis of this system is described in section 16.6.2. The annual subsistence crops grown include corn, yams, and cassava. The development of stands of cocoa leads naturally to a reduction in cultivation frequency (counting the perennial crop area as cultivated land). Cocoa–annual crop systems in Nigeria are described by Galetti et al. (1956).

c. *Rubber and rice.* (Southeast Asia). The natural small-scale topographical variation in high-rainfall regions of Southeast Asia is exploited by small farmers growing wet rice in lowland areas and rubber on adjacent upland sites near the domicile, which is of course, on high ground. The system is well developed in Malaysia and Indonesia (Courtenay, 1965). During the establishment phase, rainfed crops such as corn, upland rice, or cassava may be planted between the young rubber trees, and even at maturity the rubber area may also include coconuts and fruit trees.

d. *Oil palms and annual crops.* (West Africa). In West Africa, irregular naturally occurring groves of oil palms are harvested. The main labor effort goes into growing subsistence annuals such as corn and cassava. Zeven (1967), quoted by Ruthenberg (1971), describes such a system in Dahomey.

e. *Coconuts and annual crops.* (South and Southeast Asia). Throughout South and Southeast Asia and Oceania, particularly in littoral areas, small farms growing coconuts for sale as copra are found. In the Pacific Islands, the cropping system is as described in section 16.5.2. The degree to which the coconut areas are cropped with annuals depends on whether soil water and soil fertility are adequate for intensive cultivation. In areas where ruminant livestock are owned, natural pastures develop under coconuts and crops are often grown elsewhere.

■ 16.7 ANNUAL CROPS IN THE ESTABLISHMENT OF FOREST TREES

■ 16.7.1 The Burmese word *taungya* is used throughout the Anglophone tropics to describe a regulated system of plantation forest tree establishment, in which a public authority allocates to shifting cultivators forest land to clear and plant in food crops, in return for which the cultivators are required to plant tree seedlings in the clearing and to tend them until the cropfield is abandoned. An effort is being made (King, 1968) to popularize the term "agrisilviculture" in place of *taungya* on the grounds that its classical etymology will make it more acceptable internationally. Essentially the system is a regulated development of shifting cultivation where the farmer is a short-term tenant or licensee of an area of land owned by a state authority.

■ 16.7.2 King (1968) has assembled an exhaustive account of *taungya* practices throughout the world through the medium of questionnaires, and Roche (1974) has reviewed the system in Nigeria. King lists seventy-nine tree species and forty-two agricultural crops in use in *taungya* systems. Teak (*Tectona grandis*) is by far the most widely planted tree. Systems vary widely in respect of the degree of control exercised and the formality of the contract: for example, in one-third of the instances recorded by King, no contract documents were drawn up. In most situations, the authority is the forest department, which concentrates its attention on the tree component while permitting normal local cropping procedures, though there may be prohibitions on certain crops. In this respect, the resulting multiple land-use pattern is not so much a careful ecological plan as a coordination of established plantation forest technology with traditional local agricultural practice.

REFERENCES

Barrau, J. 1961. *Subsistence agriculture in Polynesia and Micronesia.* Honolulu: Bernice P. Bishop Museum.

Courtenay, P. P. 1965. *Plantation agriculture.* London: Bell.

Food and Fertilizer Technology Center for the Asian and Pacific Region. 1974. *Multiple cropping systems in Taiwan.* Taipeh.

Galetti, R., Baldwin, K. D., and Dina, J. O. 1956. *Nigerian cocoa farmers.* London: Oxford University Press.

Grigg, D. B. 1974. *The agricultural systems of the world. An evolutionary approach.* London: Cambridge University Press.

King, K. F. S. 1968. *Agri-silviculture (the Taungya system).* Department of Forestry Bulletin no. 1. Ibadan, Nigeria.

Ochse, J. J., and Terra, G. J. A. 1934. Het onderzoek naar de economische en landbouwkundige toestand en het voedselverbruik te Koetowinangoen. *Landbouw.* 10:1.

Parsons, D. J. 1970. Agricultural systems. In: Jameson, J. D., ed. *Agriculture in Uganda.* 2d ed. London: Oxford University Press.

Roche, L. 1974. The practice of agri-silviculture in the tropics with special reference to Nigeria. In: *Shifting cultivation and soil conservation in Africa.* FAO Soils Bulletin no. 24. Rome.

Ruthenberg, H. 1971. *Farming systems in the tropics.* Oxford: Clarendon Press.

Terra, G. J. A. 1954. Mixed-garden horticulture in Java. *Malayan Journal of Tropical Geography* 3:33.

Terra, G. J. A. 1958. Farm systems in South-East Asia. *Netherlands Journal of Agricultural Science* 6:157.

Westphal, E. 1975. *Agricultural systems in Ethiopia.* Wageningen, Netherlands: Centre for Agricultural Publishing and Documentation.

Willis, J. C. 1914. *Agriculture in the tropics.* 2d ed. London: Cambridge University Press.

Zeven, A. C. 1967. *The semi-wild oil-palm and its industry in Africa.* Agricultural Research Report no. 689. Wageningen, Netherlands.

Chapter 17

The Role of Livestock in Annual Cropping Systems

■ 17.1 CLASSIFICATION OF ROLES: INTEGRATION OF LIVESTOCK AND CROPPING

■ 17.1.1 *Classification.* Livestock in annual cropping systems may fulfill a number of roles. One classification is given below:

a. Production role.
   i. For subsistence food.
   ii. For sale.
b. Investment role.
   i. As a current "buffering" investment.
   ii. As an investment for old age or incapacity, or to bequeath to offspring.
c. Sociocultural role.
   i. As an element of social prestige.
   ii. To fulfill cultural obligations.
d. Energy role.
   i. To provide draft power.
   ii. To provide domestic fuel in the form of dung.
e. Nutrient role.
   i. To provide manure and ash nutrients after burning.

■ 17.1.2 *Integration of livestock and cropping.* The relative importance of the crop and animal components in mixed tropical farming systems, and the degree of integration between them, varies widely according to physical environment and socioeconomic conditions. With respect to integration, perhaps the most highly organized patterns are found on the Indian subcontinent in areas where natural forage is in short supply. Here very finely balanced systems have evolved under which subsistence or cash crops are grown with the assistance of draft animals, whose diet comes largely from crop products and residues not consumable by man, the provision of which requires a substantial human energy input. In addition to male draft animals, a small replacement herd may be maintained and milk may be sold. In some situations, particularly under irrigation, cropland is set aside for growing forage for milk animals and draft animals.

However, over a wide range of tropical annual cropping systems where livestock are maintained, particularly rainfed cropping systems with ruminants where natural forage is plentiful, the animals are regarded merely as a means of utilizing feed not readily harvested by man and are allowed to fend largely for themselves. Little attention is paid to their productivity in terms of liveweight gain or breeding performance, and the human input of livestock husbandry is often directed more to the security of the stock than to their nutrition. They may be carefully herded by day and housed by night, but this effort is essentially one of husbanding valuable possessions.

■ 17.2 Production roles

■ 17.2.1 *Nonruminant production.* Nonruminant animal production enterprises are a widespread adjunct to tropical annual crop farming. The geographical distribution of pig-raising is much influenced by religious taboos, but avian production is almost universal, whether chickens, guinea fowl, or ducks. Ducks are associated more with high-rainfall or irrigation areas.

In general, the role of nonruminants is the minor but useful one of scavengers, for pigs and poultry can be maintained on the crop farms, admittedly at a low production level, with little effort. Normally they are expected to find a substantial proportion of their feed requirements from free-range foraging around the domicile, supplemented with grain or tuber carbohydrate, household waste and perhaps some protein from grain legumes. Their care and maintenance is often in the hands of the female or junior members of the family, and as a consequence is not highly competitive with the main crop production effort.

Two exceptions to the above have already been noted, at opposite ends of the cropping system spectrum. First, the close integration of commercial nonruminant production with intensive irrigated vegetable production, and second, the sweet potato–pig shifting cultivation of the New Guinea highlands. The latter is a rather specialized example of a general integration of tuber cropping and pig-raising throughout Oceania.

However, whatever the degree of purposiveness in the production system, nonruminants fulfill a vital function in providing animal protein for the farm family and as a source of cash income. Their nutritional role is particularly important when opportunities for supplementary hunting or fishing are limited by environment, or where human intake of plant protein is low: for example, when the main human food energy sources are noncereals such as cassava, taro, or sweet potatoes.

As sources of food or cash nonruminants have two advantages over ruminants: they are individually small and their generation cycles are short. A small farmer owning perhaps five adult cattle would not contemplate killing one merely for home consumption, and would only contemplate sale when forced by economic necessity, since to do so would reduce his animal capital by 20 percent. Furthermore, it would take 4 to 5 years to raise a replacement of the same meat yield. But killing one pig out of a herd of twenty, or two chickens or ducks out of a flock of fifty, does not represent a significant reduction in capital and herd or flock numbers are rapidly regained.

■ 17.2.2 *Ruminant production*: *meat*. The maintenance of ruminants by tropical crop farmers for meat production, that is, with the planned objective of a continuing surplus for sale or home consumption, is fairly uncommon in Asia, though less so perhaps in Africa or tropical America. If a crop farmer maintains nondraft ruminants, his reasons for doing so are more likely to be related to investment or to the achievement of social status or influence. However, if local environmental conditions favor the creation of a regional surplus and an outlet is found, production and marketing systems tend to develop, in a rather unsystematic way. A good example is northeastern Thailand, where subsistence crop farmers run cattle and buffalo on common land and from time to time sell the surplus. In this way the region has become an important if poorly organized source of draft and meat animals for the intensively cultivated Central Plain and for Bangkok.

However, whether or not the production system has evolved to the stage of organized regional supply, the slaughter of healthy large ruminants—cattle or buffalo—for home consumption is rare and is normally reserved for cultural, religious, or hospitable occasions. The lactating cow or buffalo may provide milk for the farm family or for sale, but the only ruminant meat likely to be consumed as a dietary supplement is from animals that have died, are about to die, or have become incapacitated; e.g., a draft animal with a broken leg. On the other hand, this is less true for sheep and goats: again, the individual animals are smaller and the regeneration period shorter.

■ 17.2.3 *Ruminant production: milk.* Over a wide range of tropical crop farming systems that include the maintenance of ruminant stock, lactating animals, whether buffalo, cattle, sheep, or goats, provide milk for home consumption. This is a common practice in South Asia and in Africa, but it is far less frequently met with in Southeast Asia.

On the Indian subcontinent, particularly in irrigation areas, small-scale milk production from cattle or buffalo, for home consumption and for sale, is highly integrated with crop production. Crop residues are carefully conserved and forages grown specifically for ruminant feed: e.g., millet, sorghum, or corn as forage in the wet season, berseem clover (*Trifolium alexandrinum*) in the dry season under irrigation. Some of this will be fed to draft animals. Thus in the six categories of farm surveyed by Singh and Chancellor (1975) in the Meerut district of India (see section 15.5), the percentages of total crop area sown to forage were (a), 34.5 percent; (b), 19.1 percent; (c), 25.2 percent; (d), 17.9 percent; (e), 35.5 percent; (f), 15.6 percent.

The number of ruminant animals and the areas sown to forage in the four case studies of irrigated farms in India and Pakistan discussed by Ruthenberg (1971) are given in Table 17.1.

■ 17.2.4 It has already been noted from Odend'hal's study of a region in India (1972) that total useful energy output of the cattle population was partitioned into 84 percent dung, 12 percent useful work, and 4 percent milk. The overall energetic efficiency for milk production alone was less than one percent. However, it is once again stressed that the expression of human food output in energy terms has meaning only when the product concerned is basically an energy food, which milk is not.

Although the scale on which milk is produced on Indian farms is often very small—a few liters a day—it can be a useful cash product.

Very large dairy co-operatives with a fine network of collection points down to the village level have been established. Small farmers can deliver milk to those points daily and be paid the next day, after quality tests have been run.

Such farm milk production systems must be distinguished from urban dairies, another characteristic mode of liquid milk production in India (and formerly, before fast road transport and refrigeration, in more developed nations). Here cattle or buffalo are housed and fed at

TABLE 17.1
Ruminants and forage on irrigated farms. India and Pakistan

	Bombay (millet–peanuts)	Punjab (cotton–wheat)	Uttar Pradesh (sugarcane–wheat)	Madras (tobacco–millet)
Crop area sown to forage				
Summer	0.6%	40.1%	47.1%	nil
Winter	16.3%	38.8%	29.0%	11.8%
Draft animals	4	2	2	2
Lactating animals	2	2	2	2
Total animal units (presumably including immature stock)	11.0	6.0	4.4	4.5

the locus of product consumption, i.e., in the town or city, and are maintained on fresh or dry forage or crop residues brought in from the country.

■ 17.3 Investment and sociocultural roles

■ 17.3.1 *Investment role.* In earlier chapters the investment and sociocultural roles of livestock in crop farming systems were touched upon. The animals concerned are normally ruminants, though we have noted that pig husbandry in New Guinea and Oceania generally plays a significant role in cultural patterns. Cattle, buffalo, sheep, or goats may represent a current buffering investment, like a bank account, being added to in good times and realized in bad, or they may be maintained as an annuity insurance against old age or incapacity, or as a heritage to be passed to one's offspring. Though with the continued expansion of cash economies in developing countries we shall presumably witness a gradual attrition of these practices, yet in nations with volatile currencies they are a more secure investment than money in the bank, insurance policies or trust funds, and in times of social insecurity—insurrection, civil war, etc.—doubly so.

The use of ruminant livestock as a form of investment often gives rise, within a given village or rural area, to greater inequality in numbers of animals owned per farm than in crop area per farm. Except in situations of catastrophe, owner-occupiers can normally retain the land they own, and their current economic circumstances may be reflected more in livestock ownership.

■ 17.3.2 *Sociocultural role.* For an agronomist to rush in where even social economists and social anthropologists sometimes fear to tread would be foolhardy, and this section will be extremely brief. We may note, however, that the accumulation of livestock can represent a visible source of wealth that commands prestige in the community; that livestock may be lent, the lender thereby achieving latent economic dominance over the borrower; and that in circumstances such loans may be foreclosed upon to achieve actual economic dominance, perhaps involving the transfer of labor, land, or other forms of wealth from the borrower to the lender.

Livestock are frequently involved in cultural obligations: as a basis for periodic festivals, as a bride price, or as sacrifices. The religious sanctity of cattle in India is, of course, the prime example of a cultural force operating in the maintenance of livestock.

■ 17.4 Energy and Nutrient Roles

■ 17.4.1 *Draft power.* The role of draft animals in providing power for the tropical crop farm has been discussed piecemeal in previous chapters on the energetics of cropping systems. The important features are recapitulated in summary form below:

a. The regional distribution of draft animals in the tropical world is much influenced by historical factors. In Asia, using ruminants for draft is an ancient practice, whereas in America draft cattle date only from post-Columbian times; in tropical Africa they have a history of less than two centuries.
b. The maintenance of draft animals, substituting for human energy in the primary task of land preparation, permits an increase in the potential area cultivated from an upper limit of about 2 ha per adult male with hoe cultivation (Haswell, 1973) to 8 to 16 ha per pair of oxen (Charreau, 1974).
c. The amount of human energy that has to be expended in

maintaining draft animals varies greatly with the "free" forage resources available, whether on the farm or on common land. In some regions, e.g., parts of Java, where draft cattle or buffalo are in wide general use, human population is so dense and land use for cropping so intense that only hoe cultivation is possible. On the other hand, for example in India, human pressure on land has resulted in the evolution of intensive systems of draft animal feeding on crop residues, etc., that may absorb up to half of the crop farmer's total work energy expenditure (Boserup, 1965).

d. In general, draft animals are used only for land preparation, transport, and for stationary farm machines. Perhaps the most important of the last-named tasks is the lifting of water from wells and streams; other tasks include cereal threshing, crushing of sugarcane, etc. The evolution of draft animal implements for within-crop cultivation has been limited, though current research and development are active.

e. Work time per day from cattle and buffalo is limited to about six hours, and draft animals are rarely worked in the heat of the day. Particular care must be taken with buffalo. In seasonal climates, forage supply may at times be so short that draft animals can only be called upon for light tasks, though fortunately such periods normally coincide with the slack phases of crop farming.

f. Energy output/input ratios for draft animal cropping systems calculated on the basis of human energy input alone may be four times as great as those calculated on human plus animal energy input. However, where the sources of food energy for the draft animal are not competitive with sources of food energy for humans (i.e., forage is not grown on cropland, nor human crop products fed to stock), a more appropriate comparison is obtained by contrasting the human energy inputs for crop production alone with those for crop production plus animal care and feeding. As already stated, the latter may exceed the former by anything up to 50 percent. Where natural forage is abundant, the additional human energy expended in caring for draft stock is rendered less competitive with energy expended in crop production by delegating animal care to the younger and weaker members of the farm family.

■ 17.4.2 *Dung as fertilizer and fuel.* Again, since the functions of farm animals as providers of fertilizer and fuel have been dealt with in previous chapters, all that is needed here is a recapitulation:

a. The widespread use of animal manure as a source of nutrients for the main crops of the farm is largely restricted to situations where human pressure on arable land is such that it is cropped every year with little or no restorative fallow phase.
b. However, well before this situation is reached, the inevitable accumulation of animal manure resulting from animals being housed at night in or near the domicile is utilized on gardens and fields close at hand. With diminishing crop yield responses to increasing nutrient supply, we would expect maximum farm yield response to a given amount of manure if it were allocated more or less evenly over the whole crop area, or at least over the nonlegume crop area. On the other hand, the task of applying it to fields is energetically less demanding if it is used close to home. The actual distribution pattern within the farm will depend on the resolution of these opposing factors.
c. In tropical farming, animal manure is highly variable in nutrient quality, which is dependent on the nutrient content of animal intake, on the efficiency of retention of nutrients by the animal, and on the condition under which the manure is "made." Manure from animals maintained largely on low-quality roughage, probably through evolution genetically efficient in the digestion of nutrients (especially nitrogen), and leached in storage by rainfall, will be of low quality. High-quality manure is likely to be obtained only from intensive nonruminant or dairy production.
d. Animal manure has only a transient effect on the organic matter content and physical structure of tropical soils under intensive cultivation, and its value lies almost wholly in its nutrient content, though continued annual application of manure can have favorable physical effects.
e. Dung comprises a high proportion of the energy output of

ruminants, and in situations where wood is in short supply and economic circumstances limit the purchase of alternative fuels, it may be utilized domestically for cooking and firewood. Losses of nitrogen and sulfur from the farm system by burning dung can be substantial, but other elements retained in the ash may be returned to crops as fertilizer.

REFERENCES

Boserup, E. 1965. *The conditions of agricultural growth.* Chicago: Aldine.

Charreau, C. 1974. *Systems of cropping in the dry tropical zone of West Africa with special reference to Senegal.* Hyderabad, India: International Workshop on Farming Systems. ICRISAT.

Haswell, M. R. 1973. *Tropical farming economics.* London: Longman.

Odend'hal, S. 1972. Energetics of Indian cattle in their environment. *Human Ecology* 1:3.

Ruthenberg, H. 1971. *Farming systems in the tropics.* Oxford: Clarendon Press.

Singh, G., and Chancellor, W. 1975. Energy inputs and agricultural production under various regimes of mechanization in northern India. *Transactions of the American Society of Agricultural Engineers* 18:252.

Chapter 18

Tropical Farming Systems Research

■ 18.1 THE MEANING OF "FARMING SYSTEMS RESEARCH"

■ 18.1.1 In this book an attempt has been made to describe pattern and process in tropical farming systems as they exist, with only passing reference to the technological advances being made that will change them in the future. This final chapter will be concerned with the future, but not with the technological advances themselves, which, since progress is being made on so many fronts, would result in a heterogeneous summary of research findings. The subject is the general mode of approach to farming systems research in the tropics.

■ 18.1.2 Unfortunately the term "farming systems research" has become partially debased. The phrase was attached originally and appropriately to the multidisciplinary study of the biological, physical, economic, and social components of farming systems. The lowest common denominator of "farming systems research" is now, however, virtually indistinguishable from straightforward agronomy, which is an essential component of the larger entity, but only a component.

On the other hand, it must be recognized that, in different circumstances and in different institutions, the term farming systems research may have valid broad and narrow connotations. In its narrowest sense, the term may be applied appropriately to agronomic and field physiological research into crop combinations, both of sequences extending over more than one season and of intercropping, relay cropping, and sequential cropping situations. A more precise description of this type of investigation is "cropping systems research" and the term is in general use (Harwood, 1974). The work of Krantz and his associates at ICRISAT

(1974) is an excellent example, but it must be stressed that this work is only one part of a much broader farming systems research program at ICRISAT.

■ 18.2 THE OBJECTIVES OF FARMING SYSTEMS RESEARCH

■ 18.2.1 Before we attempt to analyze farming systems research in the broad sense, the objectives of an integrated farming systems research program must be defined. Two major tropical agricultural research institutions with designated farming systems research units have defined within a small compass the objectives of their programs. These summaries serve as a useful base for further examination.

First, ICRISAT at Hyderabad, India (Krantz, 1974):

a. To provide economically viable, labor-intensive technology for improving and utilizing the productive potential of natural resources while maintaining the quality of the environment.
b. To develop superior land and water management systems which can be implemented and maintained during the extended dry season, thus providing additional employment to people and better utilization of available animal power.
c. To contribute to raising the economic status and the quality of life of the people in the semi-arid tropics by developing farming systems which increase agricultural output and make it more stable from year to year. (The semi-arid tropics are the specific zonal responsibility of ICRISAT.)

Second, IITA at Ibadan, Nigeria (Okigbo, 1974):

a. To help develop optimum systems of cropping for long-term sustained yields for the humid tropics. (The humid tropics are the specific zonal responsibility of IITA.)
b. To help identify adapted crop varieties and determine their yield potentials in continuously cropped tropical soils.
c. To help find solutions to problems of soil fertility, chemistry, and management resulting from continuous intensive cropping as compared with traditional methods.
d. To aid identification of pest and disease limitations under traditional and intensified cropping and develop control methods.

e. To develop mechanical techniques for upland agriculture suitable for continuous cropping systems including tillage, spraying, weeding, harvesting, and drying equipment with emphasis on smallholders' problems.

f. To identify optimum combinations of crops and ways of achieving maximum returns from land under scientific management and traditional systems.

g. To examine practices and social organization of farmers with a view to identifying the major constraints in change.

The list concludes with the following: "Briefly stated, the objective of the IITA Farming Systems Program is the development of a range of efficient packages of technology for maximising production, that are economically viable, adaptable and acceptable to farmers in different parts of the humid tropics."

■ 18.2.2 It is rather difficult to compare these two sets of objectives since they are expressed at different levels of generality. That of IITA is analytic and specifies a series of linked approaches; that of ICRISAT endeavors to express ultimate goals in a more holistic manner. Furthermore, while some facets specific to the institutions concerned have been identified, there are more that have some degree of specificity: e.g., the emphasis on animal power at ICRISAT, which is in India, and the contrasting emphasis on mechanical power at IITA, in Africa, where it is expected that advances from hoe cultivation will bypass the animal power phase.

Nevertheless, the two sets of objectives reveal basic common features. In both cases, the expressed aim is to develop, through research, transferable technology that is at once production-generating, resource-conserving, and economically and socially acceptable to the farming community. But, it may be countered, is this not the basic aim of applied agricultural research in general? This is quite true, and perhaps the one criterion that is distinctive to farming systems research *sensu amplis* is an organizational one: that research in any one of these areas of increasing production, conserving resources, increasing economic return and stability, and improving social well-being—research that is necessarily carried out by different individuals or groups—is performed cooperatively and with a consistent eye to its interactions with the other areas.

■ 18.3 The components of farming systems research

■ 18.3.1 We have already established agronomic research into crop combinations and sequences as one of the core fields of farming systems research. What other disciplinary fields are essential to an effective broad-scale farming systems research program? We must beware of being too inclusive, for conceptually we can shelter virtually all aspects of applied agricultural research under the farming systems umbrella. This would, of course, effectively remove the necessity for any special designation of the research field.

■ 18.3.2 First we must include those research fields immediately adjacent to agronomy in the biological and physical sense, and those concerned with resource conservation. These comprise land and water management, soil fertility research, and applied soil physics, together with the broader fields of climatology and hydrology. Collectively, research in these discipline areas, coordinated together and with agronomy, takes care of the physical and biological aspects of crop-environment relations.

(It is appropriate to mention here that agronomic research should center not only on crops but also on their unwanted competitors, weeds. Whether the other disciplines of plant protection—crop pathology and entomology—should be included in farming systems research units is debatable and must remain largely a question of institutional organization.)

Second, we must include those research fields concerned with the economic and social aspects of existing farming systems, with the economic and social consequences of new biological and physical technology, and with economic and social restraints to the acceptance of new technology. These include production, marketing, and social economics.

The third essential research field is agricultural engineering, concerned with that much-neglected area, the actual tools by which the objectives of the farmer and his work force are translated into action. Agricultural engineering does not necessarily imply mechanization: in the developing tropics the agricultural engineer is perhaps likely to be concerned with the design and operation of hand tools and implements for draft animals or with human and draft animal energetics.

The above cast for a farming systems research group has been identified in terms of established research fields. Scientists in one field have specific methods of approach to research which differ from those in other fields. However, our assessment would be incomplete without mentioning a mode of approach essential to any multidisciplinary study of complex systems: that is, systems analysis. This may be catered for by the appointment of specialists in the field, ready to apply their skills to agronomic, hydrologic, economic data, etc.; alternatively, systems analysis techniques become part and parcel of the methodological armory of the individual agronomist, hydrologist, economist, etc.

■ 18.3.3 Without making an attempt to assign a specific weight to any individual field, we may therefore postulate the following discipline sectors and subsectors as potential components of an inclusive farming systems research program:

a. Agronomy: (i) crops; (ii) weeds; (iii) cropping systems.
b. Soil science: (i) applied soil physics; (ii) applied soil chemistry.
c. Crop-environment relations: (i) agricultural climatology; (ii) agricultural hydrology.
d. Physical management: (i) land and water management; (ii) agricultural engineering.
e. Economics: (i) production economics; (ii) marketing economics; (iii) social economics.

■ 18.4 Evaluation of established farming systems and of environments

■ 18.4.1 In section 18.3.2 reference was made to the study of the economic and social aspects of existing farming systems as a necessary activity in a farming systems research group. This is, however, only part of the evaluation of actual farming practice that is needed if new technology is to be economically and effectively generated and transferred. It is almost self-evident that if a research unit hopes to modify or transform agricultural practice in a specific region or environment, it is likely to be successful only if it operates in the full knowledge of current practices and, if it can be elucidated, of the rationale of such practices. This involves more than economic and social surveying and subsequent data codification and analysis; it requires an analytic study of farming operations and of the natural resources available for exploitation by the farming community concerned.

One of the more vital components of resource survey and analysis is agroclimatological research, aimed at classifying environments climatically in terms that are relevant to farming systems technology. A complementary activity is the description and systematization of environments in hydrological, soil physical, and soil chemical terms. The regional evaluation of weed incidence and pest and disease complexes, where this is feasible, is another example.

■ 18.4.2 The survey and analysis of natural resources forms a bridge between two types of research activity. On the one hand it is an essential complement to the agronomic, economic, and social evaluation of farming systems as they exist, and thus provides the background for relevant research. On the other hand it forms a basis for the effective transfer of biological and physical technology to a wide range of environments.

The role of physical and biological resource evaluation and analysis in the transfer of technology is an economizing one. Owing to financial and practical limitations, field research on the components of farming systems can be carried out only at a limited number of sites. If the physical environments to which research results are intended to apply are understood and ordered in a manner that is relevant to farming systems operations, then the type locations at which research is conducted can be chosen more rationally and economically; hence, extrapolation of results to different environments becomes more reliable.

■ 18.5 Evaluation and transfer of new technology

The central problem of farming systems research and extension is the evaluation and transfer of new technology. Some of those concerned with agricultural research and extension in the tropics have developed a high degree of confidence in the capacity for new technology to be transferred over a range of environments as a result of the success of CIMMYT and IRRI with wheat and rice. It is no reflection on the magnificent efforts of these institutions to say that the creation, testing, and dissemination of new crop genotypes, and of the associated crop-specific package of production technology, is, compared with the generation and transfer of farming systems technology in the broad sense, relatively straightforward. This is particularly true with respect to irrigated wheat and rice, since the degree of human control that can be exerted over the crop's immediate environment diminishes the background noise of natural physical and biological variation.

Farming systems technology—for example, new cropping patterns, new field and farm layouts, new combinations of implements and field geometry, particularly if directed toward rainfed crop agriculture—is far less readily transferable, and the bigger the "package" or complex of practices that is to be transferred the more difficult the task of transferring it. The modifications to farming systems research that have to be made to counter these limitations include an extended network of regional evaluation, an emphasis on the testing of packages or components thereof in the real-life farm situation, and a search for components of improved technology that may be transferred singly rather than of necessity in combination with others. As already stressed, the process of transfer is greatly facilitated by a codified understanding of existing farm practices and of the biological, physical, social, and economic environments toward which the transfer is directed.

REFERENCES

Harwood, R. R. 1974. *Resource utilization approach to cropping systems improvement.* Hyderabad, India: International Workshop on Farming Systems. ICRISAT.

Krantz, B. A., and Associates. 1974. *Cropping patterns for increasing and stabilizing agricultural production in the semi-arid tropics.* Hyderabad, India: International Workshop on Farming Systems. ICRISAT.

Okigbo, B. N. 1974. *The IITA farming systems program.* Hyderabad, India: International Workshop on Farming Systems. ICRISAT.

Index